Weather modification:
prospects and problems

Weather modification: prospects and problems

GEORG BREUER

TRANSLATED BY HANS MÖRTH
DEPUTY DIRECTOR, CLIMATIC RESEARCH UNIT
UNIVERSITY OF EAST ANGLIA

CAMBRIDGE UNIVERSITY PRESS

CAMBRIDGE
LONDON   NEW YORK   NEW ROCHELLE
MELBOURNE   SYDNEY

Published by the Press Syndicate of the University of Cambridge
The Pitt Building, Trumpington Street, Cambridge CB2 1RP
32 East 57th Street, New York, NY 10022, USA
296 Beaconsfield Parade, Middle Park, Melbourne 3206, Australia

© 1976 Deutsche Verlags-Anstalt GmbH, Stuttgart

English edition © Cambridge University Press 1980

First published in German as *Wetter nach Wunsch?* by
Deutsche Verlags-Anstalt GmbH, Stuttgart 1976
English edition first published by
Cambridge University Press 1980

Printed in Great Britain at the
University Press, Cambridge

Library of Congress Cataloguing in Publication Data
Breuer, Georg.
Weather modification.
Translation of Wetter nach Wunsch?
Includes bibliographical references.
1. Weather control. I. Title.
QC928.B7313   551.6′8   78-73236
ISBN 0 521 22453 5
ISBN 0 521 29577 7 pbk.

Contents

88322

Weather modification:
prospects and problems

In the late 1960s, a group of American barley growers in the San
Luis Valley of Colorado decided to enlist the services of commercial
weather consultants to assist with the control of weather for the
production of an optimum premium-grade Moravian barley grown
under contract for a brewery. The required weather modification
included supplementation of rainfall during the early part of the
growing season, suppression of hail in mid-summer and suppression
of rain during the final stages of ripening.

Soon thereafter a vociferous opposition group formed, claiming
that cloud seeding had reduced the overall rainfall in the region, and
adversely affected the growing of lettuce and potatoes. Appeals were
made to the governor and other officials of the state.

A new state law, regulating the application of weather modification
projects, was eventually drawn up and came into effect during the
summer of 1972. The leader of the aforementioned opposition group
was appointed member of a state advisory committee which decided
on applications for permission to conduct weather modification. At
a public hearing of the request for a permit under the new law, the
opposing farmers turned up in large numbers. The advisory
committee recommended that the permit be granted – and it was.
Several weeks later a bomb damaged a trailer of the firm of weather
consultants. By mid November of that year there had still been no
arrest made in connection with the bombing. A straw vote taken at
a ballot on 7 November produced a 4 to 1 vote against modification
of weather in the valley.

In the original German version of this book (*Wetter nach Wunsch*?)
I had to explain at this point that these opening paragraphs are not
taken from a science fiction novel or from the scenario of a
futurologist, but from the report of an American sociologist on public
reaction to weather modification,[1] for in central Europe such activities
are as yet practically unknown. In America, however, events like

those in San Luis Valley are just one example from many others. The ferocity of this controversy resulted largely from the brewery having applied pressure on the growers: 'Unless we have good assurance of an adequate weather management program for the barley growing season,' they had written to them, 'the Moravian barley allotment must be reduced by 20 per cent.' Further cuts would follow in latter years if no weather modification were introduced, for the company was not willing to place itself 'at the mercy of the natural elements which, in our experience, will give us a good quality crop only once in 20 years'.[2]

Commercial weather modification has been practised in the USA since the late forties. Quite a number of firms claim to be able, within certain limits, to change the weather according to the requirements of their clients. The total turnover of these firms is now about one to two million dollars a year. Among their clients are farmers from all parts of the USA, plantation owners from Latin America, Africa and Asia, and many public utilities; the Southern California Edison Company has been operating a seeding programme continually since 1950 to augment precipitation in the catchment area of their hydro-electric power station in the San Joaquin Valley. The City of New York also called on the rain makers for help when the town's water supply was in a critical situation in 1949/50.[3]

The United States' Federal budget funds allocated to research and development in the field of weather modification, rose from 2.7 million dollars in the fiscal year of 1963 to a peak of about 20 million in 1973. Because of the general reduction of research funds, the amount fell again to about 14 million in 1977.[4] Also, there is probably considerable hidden spending by the Department of Defense, on classified activities in this domain.[5] In 1973, commercial and other operational weather modification activities covered about 4% of the area of the United States, equal to about one and a half times the area of the United Kindom. Altogether, there were 67 projects, of which 55 were operated by commercial firms, 6 by municipal districts, 2 by universities and 4 by others. Among the clients were airport authorities, airlines, the Port of Seattle (fog dispersal), public utilities, water district authorities (precipitation increase), farmers' associations and individual farmers (hail prevention, rain enhancement), counties and other public bodies (mostly orders in support

viii

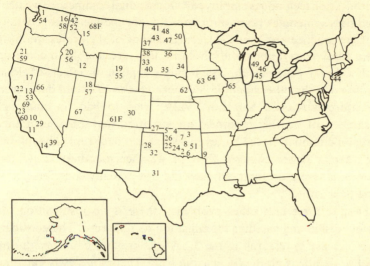

Fig. 1. Reported federally and nonfederally sponsored weather modification activities in the USA – 1 November 1972 to 31 December 1973. Numbers indicate location of projects in the order that they were reported. An 'F' indicates a federally sponsored activity (after M. T. Charak and M. T. DiGuilian *op. cit.*).

of agriculture). Lawrence County Winter Sports Inc. in South Dakota ordered snowpack increase for February and March 1973.[6]

In the USSR, operations for hail protection have become routine in some agricultural regions totalling about 5 million hectares, more than twice the area of Wales. Fog dispersal is operational at several Russian airports and there are some experimental schemes for increasing precipitation.[7] Australia and Israel have a long record of rainfall stimulation experiments. An operational programme was due to start in Israel recently. In France, fog dispersal is in operation at the Orly and Charles de Gaulle airports near Paris, after a long period of pioneering experiments.[8]

The theoretical foundation, on which weather modifiers based their first activities shortly after the end of World War Two, was rather scanty. To date, there is only one kind of weather modification which is reliable beyond doubt and is generally acknowledged: the dispersal of fog over limited areas, such as airfields. There is still no general agreement in professional meteorological circles as to whether and to what extent one can augment precipitation or prevent lightning, hail and storms, nor regarding the efficacy of the modification techniques employed.[9] Operational commercial weather modifiers sometimes point out cynically that the methods generally work in practice and that the learned friends of the profession have their many reservations, only because they are unable to come up with a theory which would convincingly explain the effects of cloud seeding. On the other hand, it is a fact that the 1971 hail damage in the afore-mentioned San Luis Valley was so bad, in spite – or because? – of weather modification activities, that a hail insurance company founded by the barley growers went bankrupt. The following year, a newly engaged firm of weather modifiers could not prevent a severe drought – if they did not actually contribute to it – by their efforts.[10]

Despite these and other problems, there is no justification for simply dismissing weather modification as nonsense. The considerable increase in research in the USA and in other countries during the last decade is gradually bearing fruit. Theory is slowly catching up with the far advanced practice. The futile and unpleasant earlier arguments between superoptimistic commercial weather men and some extremely sceptical members of the meteorological profession who doubt that weather modification is at all possible are giving way to factual investigations into the reasons for success or failure *under given circumstances*. Nowadays scientists can answer such questions fairly accurately in many cases.

This book intends to give to the reader a balanced picture of the present state and perspectives of weather modification. It relates the successes so far, and the difficulties which have still to be tackled. It presents the arguments of those scientists who, after many years of work in this field, support the expectation that, over the next ten to twenty years, progress in rainfall stimulation and, possibly, hail suppression will be just as impressive as it has been in the case of fog dispersal. Equal regard is given to the sceptics who consider such expectations ill-founded and exaggerated. There are not only quotations from optimists who think that weather modification is a good and useful thing but also references both to voices of doubt and reservation and to those who ask if weather modification is at all desirable.

One thing needs to be said with emphasis right at the beginning: Despite the many difficulties, reservations and doubts, weather modification appears to be capable of development. The time when a meteorologist was ostracised by this colleagues for his association with this 'non-scientific' subject is quickly running out. Already, highly qualified experts are emerging, who specialise in problems of weather modification. Expenditure on research in this field is likely to grow – the result not least of pressures from the military, who have shown a great interest in this subject right from the start, and who have already applied weather modification techniques during the Vietnam war (see p. 71).

The increased research effort is likely to produce results. These may emerge more slowly than the optimists expect, but we must anticipate that development of weather modification techniques is going to proceed, regardless of theoretical research. Here, a prognostic technology assessment would be of great importance. But so far, little has been done and the meagre resources which are available for this purpose often remain unused because no scientists can be found to take on the relevant work.[11] Thus, there is a real danger that, in ten or twenty years, we may find ourselves with established techniques, the negative consequences of which will only be realised once it is too late to abandon them.

There is no doubt that negative and undesirable consequences could arise. Their assessment should not be left to the meteorologists alone, since modification of weather and climate also involves far-reaching effects on health, ecology and economics. Therefore, it would be most desirable for representatives of other disciplines – doctors, biologists, ecologists, environmentalists, economists, development planners, communication experts, sociologists, lawyers and experts on international law, system analysts and others – to look into these questions and to identify some high risk areas which are perhaps not easily recognised by the meteorologist. Now is the time for trying to foresee and avoid damaging developments. Later, when consequences are beginning to show, it will be difficult, if not impossible, to rectify matters.

It is equally important that the general public is informed of these problems as soon as they are recognised. Changes in weather affect literally everyone. It is therefore reasonable to expect the general public to want to have a say in these matters. Of course, there is the danger that, by inclusion of a large and non-specialist public body, the arguments will be judged according to the loquacity and oratorical skill of their proponents, rather than by their factual weight. But, on the other hand, restriction of the discussion to a small circle of experts and technocrats would, judging by experience, only facilitate lobbying by various groups and there would be no guarantee that the wishes and interests of the general public would be optimally satisfied.

It will not be easy to formulate these wishes and interests. Experience gained so far clearly shows that different individuals and groups of people – not only in the San Luis Valley – frequently have quite different interests in the weather. These different outlooks could conceivably become election campaign issues in the next century. Already now, there are occasions when disputes of this kind are brought before the courts. The larger the scale of weather modification, the more people will be concerned, and the more incisive will become the diverging interests. It will be necessary to create, as soon as possible, generally accepted national and international rules which can be applied to settle these disputes. These must meet the interests of all concerned, otherwise there is a danger of weather modification, even for peaceful purposes, becoming more damaging than useful and a permanent cause of friction between individuals and states.

Even more serious are the dangers which arise from the military use of these new techniques. There is no detailed information on the extent of weather and environmental weapons which are at present at the disposal of the large powers. Some scientists have pointed out the possibility of military weather modification becoming capable, in the near future, of altering the tracks of hurricanes, of triggering devastating droughts, and of modifying, with grave consequences, the upper layers of the atmosphere.[12] These dangers are intended to be contained by the Treaty on Weather and Environmental Warfare, which was signed within the framework of the Geneva Arms Limitation Talks in the spring of 1977, after years of negotiations.

A proper evaluation and application of techniques of large-scale weather modification, which will probably be possible in a few decades, would require world-wide planning and co-operation. The ban on military uses of weather modification should therefore be supplemented by an international agreement, according to which all peaceful activities in this field should proceed under the control of an international agency. Conceivably the World Meteorological Organization (WMO) could take on this responsibility. If such an approach on an international basis does not come about, one must fear that peaceful weather modification will be pursued on narrow-minded national lines, without regard to smaller and less influential nations, and thus become the object of mutual accusations and conflicts.

In preparing this text I had the support of many distinguished scientists working in the field of meterology and/or weather modification. My thanks go to Professor Dr Konrad Cehak of the University of Vienna for general meteorological advice. For answers to specific questions and information updating the text to the state of affairs at the end of 1977 I am grateful to Dr B. Federer of the Laboratorium für Atmosphärenphysik, ETH Zürich, Director of the Swiss 'Grossversuch IV' on hail prevention; Dr H. Willeke of the Deutsche Forschungs- und Versuchsanstalt für Luft- und Raumfahrt, Oberpfaffenhofen; Director E. Sauvalle, Aëroporte de Paris-Orly; Professor Dr J. Neumann, The Hebrew University of Jerusalem; Dr E. J. Smith, CSIRO, Australia; Professor Joanne Simpson, University of Virginia; Dr R. W. Sanborn, Deputy Director of the US National Hail Research Experiment, National Center for Atmospheric Research, Boulder, Colorado; Professor L. O. Grant, Colorado State University; Professor A. S. Dennis, South Dakota School of Mines and Technology; and Professor W. R. D. Sewell, University of Victoria, Canada.

I am indebted to Cambridge University Press for presenting my book to the English-speaking public and to Dr Hans Mörth of the University of East Anglia for advice and for translating my text into plain English intelligible to the ordinary reader.

The conclusions and recommendations in the last chapter are solely my own.

Vienna, May 1978   GEORG BREUER

1
Scientific and technical background

Can meteorology become an exact science?

Successful space missions of recent years vouch for the possibility of exact calculation and detailed prediction of the movement of satellites, moon rockets and interplanetary space sondes. Corrections which have to be made to the orbits of these vehicles are minute. They result from technical imperfections during launching and from the fact that it is not possible to account for all the minor variations in interplanetary forces acting on the space vehicle. However, the basic theory on which the orbit calculations depend is flawless.

There is no equally reliable theory which could be applied to the problems of weather modification. Nobody knows for sure what the exact results of a planned modification experiment will be. More often than not, it cannot even be proved that the observed developments would not have taken place anyway, regardless of the experiment. Thus, the art of 'weather making' proceeds in a theoretical twilight. An assessment of its future development is closely linked with the prospect of an adequate scientific foundation being established, and also with the question of development prospects for meteorology as a whole.

We know from daily experience that the theory of weather has not yet reached the standard of perfection of the theory of the movement of celestial bodies. Astronomers can calculate the position of planets tens and hundreds of years ahead and predict eclipses to an amazing degree of accuracy. Meteorologists are not always successful with their 24-hour forecasts, let alone with prognostications for periods of two days and longer. But is not weather also subject to physical laws? and should it not be possible to predict its evolution exactly, as long as the initial state can be determined with sufficient accuracy? It is the secret hope of many a meteorologist that the atmospheric sciences will, one day, also attain a degree of exactitude comparable to that reached by celestial mechanics several centuries ago.

1

If we were to judge the maturity of a branch of science only by the reliability of the prognostication derived from it, we would have to infer that meteorology is still awaiting the birth of its Kepler, and that even the priests of ancient Babylon were able to predict the positions of stars with a higher accuracy than present-day meteorologists forecast the weather. On the other hand, if we take the view that the primary requirement of science is the *explanation* of observed phenomena, then meteorology is relatively far advanced. Kepler summarised the observational material empirically in mathematical formulae that allowed fairly correct predictions of planetary movement to be made, but he could not explain *why* the planets moved as they do. The meteorologist of our times understands and appreciates a much wider and much more complex range of physical processes which exert control on the weather, yet he is unable to predict its long-term evolution. This must be taken as convincing evidence for weather being the end product of a greater number of more intricate processes than is the motion of the planets.

Astronomers and space mission scientists wishing to predict movement within our solar system can limit their considerations to mutual gravitational interaction between the sun and a relatively small number of planets and their satellites, all of known masses and orbital character. The meteorologist is confronted with a multitude of ever-changing air masses which interact, mix and dissolve in each other, only to be regenerated and newly contrasted in never-ending succession. Not only must he, like the astronomer, consider basic physical laws regarding atmospheric motion, conservation of energy, mass and angular momentum, but also the hydrodynamic and thermodynamic behaviour of gases, the effects of radiation, and a multitude of feedbacks and lagged interactions between atmosphere, ocean and ice sheets. 'Anybody familiar with these physical laws also knows that they involve processes, the precise formulation of which has either not yet been achieved or is too complicated for practical application', says an expert in theoretical meteorology.[1]

The earth is, to a good approximation, a rotating sphere. Its rotation exerts a planetary control on circulations in the atmosphere and oceans. However, the free flow of cold and warm ocean currents, which have great influence on weather and climate, is limited by the continents. Land surfaces greatly restrict the freedom of atmospheric motion through friction at the ground, to a degree that depends on the characteristics of the land surface. Different geographical influences in different regions of the earth complicate the formulation of generally valid empirical rules in meteorology. The Arctic is an ice sheet on an ocean surrounded by continents, the Antarctic a glacier-covered continental plateau surrounded by oceans. The larger part of the middle and high latitudes are continental in the northern hemisphere, while the corresponding latitudes in the southern hemisphere are covered by oceans. The principle chains of high mountains run in an east–west direction across Europe and Asia and form a climatic barrier between colder and warmer regions. In America, the orientation is from north to south and presents no obstacle to the meridional advancement of polar and tropical air masses. Accordingly, the weather in America is punctuated by greater and faster contrasts than in Europe. The Americans say: 'If you don't like our weather, just wait five minutes'.

Practially all energy on earth, including that which drives the atmosphere, comes from the sun. Solar radiation is the more effective the higher the sun stands in the sky. Because of the daily rotation of the earth, there is a periodic warming during the morning until early afternoon, followed by a cooling through heat radiation loss to space, which persists throughout the night until the sun rises again. The inclination of the earth's axis of rotation to the orbital plane produces the seasonal variations in maximum solar altitude and in the length of the day. These variations are reflected by dry and rainy seasons in the tropics, by spring, summer, autumn and winter in middle latitudes, and by the half-yearly duration of midnight sun and perpetual night near the poles. The zonal differences in the daily and seasonal supplies of sunlight and energy cause differences in the temperature of water and air which lead to horizontal and vertical motion in the oceans and in the atmosphere. Also, the amounts of evaporation show a strong meridional gradient. Relatively little energy is transferred to the atmosphere by direct absorption of the sun's rays. A considerable amount of heat is supplied to the atmosphere through condensation (fog and cloud formation) of water vapour, which evaporates off the oceans. Another significant portion of the heat content of the atmosphere derives from absorption of heat re-radiated by the warmed surface of the earth. The heat

3

capacity and the reflective character (albedo) of the earth's surface varies greatly with the vegetative cover and the state of the ground. Oceans, forests and land under cultivation absorb much solar energy, storing and radiating large amounts of heat. Deserts, snow, and ice reflect most solar radiation back into space but they store and transmit little heat.

All these climatic factors vary in a determined annual rhythm which explains the seasonal weather variations. But since weather is also subject to changes from year to year, there must be other influencing factors which are not entirely controlled by the annual rhythm of the earth's orbit. Among these are the 11-year sunspot cycle and variations of solar activity which affect electric and magnetic processes in the higher atmosphere. These appear to extend an influence to the lower layers of the atmosphere and our weather, but no plausible physical explanation has yet come forward (the connection between weather and solar events will be subject to an international study by a Special Committee for Solar-Terrestrial Physics in the forthcoming years). Another control may result from the input of volcanic dust and gases into the air, whereby the radiational balance of the atmosphere can be changed. Also, there are tides in the atmosphere, similar to the tides in the oceans produced by the sun and the moon. We might speculate that the positions of celestial bodies influence the weather on earth, although this idea would be hotly disputed by some atmospheric scientists, since there has been no conclusive evidence in its favour.

Finally, the present state of technology has already led to inadvertent weather modification by man and it must be assumed that climate modification, albeit local and relatively small, has taken place during past millennia as a consequence of human changes in land use and vegetative cover on the earth. There probably exist a number of other, non-cyclic factors which have not yet been scientifically established.

To make a reliable prediction, it is not necessary to be able to understand the ultimate intricacies and causality of a physical process, as the successful application of Kepler's laws without comprehension of the concept of gravitation has demonstrated. Often a well-prepared mathematical formulation of observational facts and empirical deductions is sufficient for the purpose. If the regional distribution of yesterday's and today's weather is known, and if there is empirically acquired knowledge of the behaviour and movement of air masses and fronts, a scientifically based prediction can be made without there being detailed answers to the basic questions as to why different air masses and particular frontal positions come about in the first place. But here two main difficulties confront the meteorologist.

First, the available data are still inadequate. Experience in recent decades has shown that atmospheric measurements made at the surface of the earth are not sufficient. They must be supplemented by measurements at upper levels in the atmosphere. This is done by sending up balloons with radiosondes attached – automatic recording instruments which send the data to a ground receiving station by a small radio transmitter. These radiosondes are released twice a day, at 0 and 12 hours Greenwich Mean Time, from many points on the earth. Naturally, the density of this upper air sounding network is low over oceanic and polar regions. Meteorologists depend partly on information from ships and aircraft which constitute a supplementary network that is variable in time and space. Even so, the global weather charts on which analyses and forecasts are based, still show some blank areas.

The second problem arises from the complexity of the atmospheric processes. If all presently known physical factors in the evolution of weather are considered, the resulting physical models and mathematical relationships are so complicated that even the largest and fastest available data processing cannot keep up with the 'real-time' monitoring of the weather and its prognostication. Thus meteorologists cannot find out until some time after the event whether or not their forecast follows the actual evolution – which is useful for the verification of the theories and models, but of little practical value to the user of the forecast. To keep up with time, we must therefore simplify the models and trim down the factors at the cost of accuracy and reliability.

The use of computers has enabled a much faster, more comprehensive and improved weather prognostication, but it has not yet solved the basic problem. A reasonably detailed computer forecast for the European region requires a data grid with less than 200 km distance between grid points. This amounts to about 8000 grid points on the northern hemisphere. If data for up to 20 levels from 500 to 15000 m above the surface are supplied at each of these grid points, and if the evolution of the weather is simulated by computation of the atmospheric conditions in incremental time steps of 5 minutes or, better still, 2 minutes, then 288 or 720 time steps are required for prognostication over 24 hours. Altogether, this would involve – depending on network density and length of time step – from 23 million to 115 million calculations, each of which is again composed of many single algorithmic steps. No weather service has a computer big and fast enough to cope with so many computations in a reasonably short time (1 or 2 hours).

Instead, some compromise must be found to reduce the amount of computations. Several meteorological services have developed a two-stage approach. A relatively wide mesh of grid points is used for the calculation of large-scale weather evolution in the hemisphere. Within this grid is a second, fine-meshed grid point pattern which covers only the area for which the weather forecast is issued. Experience gained in France with this method has shown that it is especially suitable for weather forecasting in mountainous regions.[2] However, as long as the distance between actual weather observation stations in many parts of the hemisphere remains substantially bigger than even the widest computer grid point mesh, grid point values can only be estimated by interpolation, and the computations are therefore not based on precise data.

A further shortcoming arises from the fact that working with grid point data introduces a certain amount of smoothing which means that small-scale atmospheric phenomena, such as showers and local thunderstorms, are being neglected. It is therefore not possible to forecast accurately when and where the sun will shine, which mountain is going to be cloud free or which will have a cloudcap, where fair weather will prevail, and which places will be affected by showers or storms. All this detail is circumscribed by typical weather forecast phrasing, like: 'fair to cloudy, local storms during the afternoon'. This is not only a serious shortcoming for the user of the forecast – a farmer, a tourist or anybody who wishes to plan his activities in a given place – but also introduces a basic uncertainty for the medium-range weather forecast (three or four days ahead), since local weather processes can influence the evolution of controlling factors considerably.

The development of weather satellites has introduced a new perspective of large-scale weather diagnosis, which could not be obtained from ground-based observations. Satellite picture transmissions allow an early detection of tropical revolving storms in remote oceanic areas, and provide the basis for accurate tracking and for warnings to shipping and coastal habitations. This enables preventive measures to be taken much earlier than was previously possible.

On 10 September 1961, hurricane 'Carla', one of the largest hurricanes in living memory to affect populated areas of North America, wreaked havoc in the Gulf of Mexico. More than 300000 people living in the coastal towns of Louisiana and Texas left the danger zone to save their lives. The evacuation of such large numbers in a timely and orderly fashion was only possible because of early warnings by the US Weather Bureau two days before the landfall of the storm. Frequently issued bulletins contained the exact positions of the storm, direction and velocity of its movement, and accurate estimates of the wind force. Never before had a meteorological event been predicted so accurately, nor was the potential loss of several thousand lives prevented so effectively. All this was as a result of weather satellite *Tiros III*, whose television cameras detected the formation of the storm much earlier than would have been possible with previous observation methods.[3]

Weather satellites nowadays supply measurements of atmospheric variables for areas lacking in ground-based observation stations, and can thereby supplement the input to the computer programmes of numerical weather forecasting. Further supplements are provided by automatic weather stations in polar regions and oceans, and by drifting balloons carrying a radiosonde-type instrument package, which transmits upper air data to any ground receiving station within range, or relays radio signals via a communication satellite. By these measures the blanks on the weather maps may be eradicated.

The use of modern technology has in recent years contributed substantially to a better understanding of the global weather and climate and has, in general, also led to an improvement of local forecasting accuracy. People whose livelihood depends on the weather – pilots, mariners and farmers – are well aware of this fact. The general public, however, has hardly taken note of it. Psychological bias plays a certain role in this. Any gross malprognosis engraves itself on peoples' memories much deeper than might several dozen fairly correct prognostications which are simply acknowledged as a matter of expectation.

We can understand the vociferous reaction over 'the incompetence of meteorologists' by people whose holiday planning efforts are frustrated by really bad weather forecasts. On the other hand, we must sympathise with meteorologists when they protest against generalisations and against being the object of never-ending cheap ridicule while their achievements are hardly ever acknowledged.

It would not be realistic to expect that weather men should, in the near future, produce forecasts of an accuracy comparable to those in astronomy. Nevertheless, a further significant improvement in weather forecasts is a realistic goal. A new European Centre for Medium Range Weather Forecasts, equipped with powerful computers, is presently under construction at Shinfield Park, near Reading, in England. It is hoped that this work will lead to a substantial improvement in the forecasting of weather for five to ten days ahead, from 1980 onwards. Also, there will be a steady development of the theories regarding the influence of non-seasonal factors on the weather, which would at least provide a more solid basis for the assessment of man's role in the modification of weather and climate.

Benefits from wrong theories

Weather modification, as practised today, is applied to small-scale local features which remain within the point distance of the computer grid and can generally not be described and forecasted in the routine weather analysis operation. It often aims at influencing a single cloud either to prevent it from producing hail or to induce it to precipitate more rain then would have occurred without human intervention. To understand the processes going on in a particular cloud, presumably we do not need to know all the factors which control the large-scale weather situation. The physical processes involved and the laws of science governing them are known in principle. However, in practice, this knowledge of the basic principles is insufficient. Each cloud has its own specific peculiarities which cannot easily be foreseen.

'The most striking lesson these writers have learned from 30 years of cloud study is that a cumulus cloud can do virtually anything all by itself without any interference by man', writes Dr Joanne Simpson, at that time head of the Institute for Experimental Meteorology, in Miami, Florida (now at the University of Virginia). 'In a field of identical-looking cumuli, one or several can explode to thundering cumulonimbus, while the rest humbly die. Given two apparently identical convective storm systems, one can rain pitchforks, while the other with indistinguishably similar looking clouds, remains dry.'[4]

Allowing for this high amount of natural variability, we must not give way to the temptation of evaluating experimental weather modification results by 'eyeball'. Many unproductive controversies have arisen in the past, Dr Simpson reports, because commercial rain makers made exaggerated claims, such as: 'I just know the seeding was a success: the seeded clouds behaved very differently from the neighboring unseeded clouds – in fact, I have just *never* seen any clouds behave the way those seeded clouds behaved.' The cloud expert will reply: 'Just stick around for a while, chum, and you'll see natural clouds behaving that way.'

Dr Simpson, who directed many scientifically well-prepared cloud modification projects herself, is not suggesting that rain makers should defer all activities until scientists can provide a perfect theoretical basis. Rather, she takes the stand that theory and practice should proceed hand in hand. An experiment would be more credible and produce more meaningful results if a 'model' existed which was able to express in mathematical form the processes occurring in a cloud. This could be fed, together with specific initial conditions, into a computer and would yield quantitative predictions of future development in seeded and unseeded clouds.

However, even modelling a single cloud raises problems similar to those encountered on a more sophisticated level in the formulation of models for numerical weather forecasting by computer. First, many of the relevant equations cannot yet be formulated exactly.

Secondly, even if an exact formulation were possible, the near future is not likely to provide a computer powerful enough to solve such equations quickly, and, thirdly, there would be no point in attempting such exactitude as long as the data available at present from instrumental measurements do not have the accuracy required for a precise definition of the initial conditions. 'Hence', concludes Dr Simpson, 'all meteorological models are, and will remain, hierarchies of approximations and simplifications. Their adequacy must be judged by the degree to which they predict the phenomenon in question; the predictions must always be checked by measurements, which in their turn are never perfect.'[5]

Well planned experiments contribute not only to the development of a theory of weather modification but also to general meteorological knowledge. On the other hand, the need for a theoretical basis of weather modification may further stimulate more intensive research of small-scale meteorological phenomena. 'The situation where a major experiment is launched without adequate fundamental research', write the weather modification experts A. Gagin and J. Neumann of Jerusalem University, 'is no exception to the rather general rule that financial support for such research becomes available (and even then on a limited scale) only when the need for applied research requires it.'[6]

Indeed, attempts at changing or exploiting natural resources without sufficient theoretical groundwork are not peculiar to weather modification. If we look hard enough, we find that a large part of human history is filled with theoretically inadequately based exploitation attempts, many of which had considerable success. The experimentalists of the grey past who bred domestic animals and crop plants had no knowledge of Mendel's laws of heredity or modern genetics. The wheel was invented without mechanics; the stone catapult and the bow and arrow were developed before the days of ballistics. Similar examples can be quoted from the history of the modern technical age. The Diesel engine was developed on the basis of a theory that was at least partly misunderstood. Edison was an inventor who made up for lack of theoretical knowledge by his diligence. On the other hand, artisans derived from their vast practical experience theories which were often faulty or irrelevant. These were then adopted by the industries which developed from

10

these crafts. An English textile scientist once started his inaugural lecture with the following statement:

No newcomer to textile research can help but be impressed with the degree of optimisation on the processes that has been achieved by trial and error methods extending over a long time. The same cannot be said for the conventional theories that have gone with these developments. One of the earliest needs of the textile engineer who wishes to modify existing processes is to disentangle the true purpose from the supposed purpose of many textile machines.[7]

The relatively short history of weather modification already seems to provide examples which show that the effects aspired to – increase of rainfall or suppression of hail – do not come about only as the result of a mechanism predicted by theory (provision of sufficient or more than sufficient numbers of icing nuclei) but, at least partly, through another originally unforeseen mechanism (intensification of the internal dynamics and of the buoyant forces of the cloud). It is probable that other theories and hypotheses which are discussed in this book will be proved faulty or of limited value in the future. Every new discovery contradicts some theoretical concepts, narrows their domain of validity, or shows them to be special cases. This applies especially to a new field such as weather modification, which has not yet secured a basis of proven knowledge, tested and confirmed by thousands of experiments. For this reason, the personal experience of a rain maker, his 'finger tip perception' which cannot be defined in scientific terms, and skill in the 'artisan' part of his work, may contribute decisively to the outcome of his operations. Of course, such an intuitive approach to the tackling of the task on hand can equally mislead.

So far, practical application of weather modification has moved ahead of theory, and experience has been accumulated which awaits interpretation by the scientists. The weather modifiers are not prepared to sit and wait. They point at the many (imagined or real) positive results of their work. It has been shown that such results can occasionally be achieved solely by observation and experiment, but, no doubt, a good theoretical foundation provides a better chance of success.

The great advances made in technology, medicine and agriculture during the last hundred years can be ascribed to a combination of 'pure' science and practical application. False or partly erroneous theories have played a positive part in these advances. Even a wrong hypothesis, as long as it is formulated precisely enough to be tested by observation or experiment, contributes more to the advancement of science than a statement to the effect that a natural phenomenon cannot be explained.

Small efforts – large effects

'To assume that a hurricane could not be successfully modified by even a single pellet of dry ice is like assuming that a very large forest could not be set on fire by such a small thing as a single match' wrote the American Nobel Prize winner Irving Langmuir, who has been called the father of scientifically based weather modification.[8] But this imaginative comparison is no substitute for scientific proof. Forest fires cannot generally be generated by a match unless the environmental conditions are suitable; and they can only be generated because a large amount of combustible matter was accumulated in the forest by the photosynthesis of plants. This may be viewed as a reservoir of chemical potential energy, which can be activated by the match. However, the instabilities required to produce a situation in which small efforts may trigger large effects do not occur all the time and everywhere in nature.

Astronomers, for example, can calculate exactly the direction, magnitude and duration of a force which would have to act on the moon to bring it 3000 km closer to the earth. They could predict with great accuracy the resultant effects on the orbital period of the moon on the tides, the earth's orbit, its rotation and so forth. Yet, with all the theoretical foundations secured, it would be impossible to realise such a project – the required supply of energy would exceed the capacity of human technology by orders of magnitude. The earth and the moon constitute a stable system which rules out the possibility of 'triggering' a large effect by a small effort.

Nevertheless, unstable situations in the movement of celestial bodies could be imagined. Let us assume that a minute planet from the asteroid belt approaches close to the earth; its orbit might be changed by a relatively small human technological effort, so that it becomes a second moon. It is probable that some of the satellites of

12

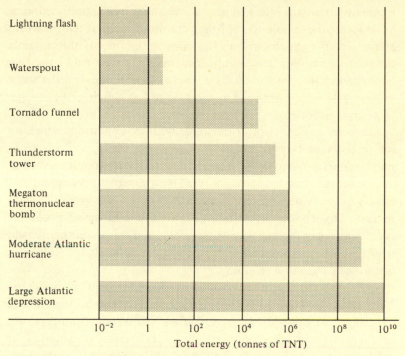

Fig. 2. Energy exchange in weather systems compared with explosive power of bombs. TNT = Trinitrotoluene, an explosive used in conventional bombs. The atom bomb that destroyed Hiroshima equalled 20000 tonnes (= 2×10^4 t) TNT, approximately the same amount of energy as contained in a tornado. (After *Science Journal* 1968.)

Jupiter, and possibly the mini-moons of Mars, are captured planetoids. Some experts even think it possible that our moon was captured in this way. However, such changes have been relatively rare in the history of the solar system.

Weather processes involve much lesser magnitudes of energy than the motion of celestial bodies. Nevertheless, the total energy transformation in the atmosphere is about a million times greater than all human power consumption. A single large storm transforms about as much energy as is liberated in a medium-sized hydrogen bomb explosion. A hurricane of moderate strength 'uses' energy approximately equivalent to that of one hundred hydrogen bombs. These are orders of magnitude which cannot be matched by human technology.[9] In fact, there is only one kind of weather modification where an attempt is made to produce changes by the application of brute force: the dissolution of fog over limited areas at temperatures above freezing. Such 'warm' fog constitutes one of those stable situations where one cannot apply a 'match' to get rid of the fog. Nevertheless the effort in terms of energy (and money) can in some instances be worth while, as in the case of fog clearing of runways of a large airport.

There are a number of other stable weather conditions which do not lend themselves to modification with presently available technical resources. Among these are – unfortunately – the long-lasting droughts in the tropics. The stability of the atmosphere is so great that clouds – even when seeded – will not grow to the vertical extent necessary to produce precipitation. Quite often, however, situations occur which can flip one way or the other, and this is another reason why reliable prognostication is much more difficult to achieve in meterology than in astronomy.

Such instability is the true hunting ground for weather modifiers. Their skill is determined by the ability to select the right 'match' and to strike it at the right place. Of primary interest, nowadays, are unstable clouds and cloud systems, which can be exploited to influence processes on a relatively small scale. Also, it is conceivable – and this has not yet been fully investigated – that a clever exploit-ation of chain reactions, feedback and amplification effects could affect the large-scale weather by way of small-scale modifications.

Langmuir is undoubtedly right when he speaks of the possible existence of meteorological 'matches', although one must not take his example of the piece of dry ice and the hurricane literally, at least not when considering planned weather modification with predictable results. For that, much stronger 'matches' are required, particularly so, since the theoretical basis for such interventions has not yet been perfected.

Small efforts – no effects

'If Christmas brings snow, the old year will soon go' is the free translation of an Austrian peasant proverb. But, what if Christmas brings none...? This facetious saying expresses a basic rule of logic: the validity of a statement of sequential events cannot be taken to prove causality.

Fig. 3. 'Little Bear to plane. Little Bear to plane. Start seeding.' (After drawing by Dana Fradon: © 1953 The New Yorker Magazine Inc.)

For thousands of years, man has appealed to higher powers during prolonged droughts. Rain-maker rituals, tribal dances and sacrifices speak for the people's desire to pacify the angry gods. Christians express the same needs by processions and prayers. Not infrequently, these appeals appear to be successful. Modern, enlightened man dismisses such 'successes' as superstition. He argues that it would have rained anyway, with or without prayer and dance. He admits that the religious ceremonies may have a psychological effect on the participants, but he does not think them to be an appropriate means of influencing the weather.

The seeding of clouds is – or at least could be – in the way of modern thinking, a suitable means of weather control. But, if we do not want to become tied up with more pseudo-scientific superstition, we must look very carefully into the success claims of weather modifiers. At first sight, some of their statistics look very impressive but, on closer scrutiny, it appears that changes observed after cloud seeding are still within the range of natural weather variations. This, obviously, does not prove that there was no actual modification of weather. There could indeed have been an increase of precipitation

15

as a result of the cloud seeding – but it cannot be conclusively shown by application of statistics.

There are certainly some examples to which this reservation does not apply. If the claim of Soviet scientist regarding a reduction of hail damage by 50 to 90% by targeted cloud seeding in different areas far apart from each other could be substantiated, and if such successes could have been repeatedly achieved over several years, there remains only the question of whether the applied effort is worthwhile. This can be answered by a cost–benefit analysis. However, rain making operations do not induce changes of such a magnitude. Their claimed successes seldom exceed 10 to 20% over long periods. Even that is a result for which the effort would be worthwhile in many cases, if we could only be sure that such precipitation increases arose as a consequence of cloud seeding rather than by chance.

Judgement on this issue can pose a grave decision-making problem, especially in small countries with a limited budget. Israel draws a large part of its water from Lake Tiberias, which yields a maximum of 350 million cubic metres a year. In the 1950s it was already clear that this amount would not be sufficient for future developments;[10] possible alternative water supplies were looked for. There was also the attractive idea of artificially reducing the evaporation, which in the case of Lake Tiberias runs to 270 million cubic metres a year. However, Australian experiments on reduction of evaporation by spraying a thin film of nonevaporating liquid on top of the water surface have failed to produce convincing results. 'Recycling', the reuse of waste water, requires considerable investment and poses serious problems of hygiene. The Israelis have a moderately sized capacity for desalination of sea water, driven partly by solar energy. Significant outputs can only be achieved with expensive nuclear plants which yield electric power as well as fresh water. If cloud seeding were a reliable method, it would be by far the best and least expensive solution.

As a result of two cloud seeding experiment projects during 1960–67 and 1968–75, the Israeli weather modifiers have come to the conclusion that precipitation over the catchment of Lake Tiberias can be increased by 10 to 15%. Statistically significant values of about 20% have been observed in the interior of the experimental area.

Scientific investigations which ran parallel with the project have shown under which specific meteorological conditions cloud seeding is likely to be successful. They have also provided a scientific explanation for the reasons. Similar experiments conducted at the same time in a region between Haifa and Jerusalem in the southern part of Israel have not yet produced entirely conclusive results and are likely to be continued.[11]

Consequently, A. Gagin and J. Neumann of the Hebrew University of Jerusalem proposed an end to the experimental phase in the northern region, and the start of an operational rainfall stimulation programme for the catchment area of Lake Tiberias. If the fruits of such an effort are to be harvested, one must augment the capacity of the pumping stations at the lake, otherwise the additional water will run off into the Jordan river. On the other hand, should the estimates of the weather manipulators prove to be overoptimistic, such augmentation would be an expensively bad investment – the lake would not yield more water than at present, and the water supply of the country would be in trouble. Israel can hardly afford to invest in several alternative costly solutions. The same applies to other arid countries. Thus the main difficulty for the decision makers arises from the confrontation of the more or less cautious optimism of the rain makers with the continuing scepticism among many meteorological experts who do not believe in reliably successful cloud seeding. The renowned British cloud physicist and meteorologist B. J. Mason puts forward this view:

Having studied developments in this field very closely for about twenty years, but without having been directly involved either personally or officially, I cannot avoid the conclusion that, with some notable exceptions, such operations have generally failed to conform to the accepted principles and standards of scientific experiment and analysis, and are therefore incapable of providing objective answers to such questions as to whether, in what circumstances, and to what extent, it is possible to modify precipitation by artificial seeding. I believe that the majority of responsible meteorologists share my concern that, in several countries, politicians and entrepreneurs, ignorant or impatient of the scientific facts and problems, are initiating and conducting major weather modification projects without the benefit of proper scientific direction, advice and criticism, and that this may have serious repercussions on the reputation of meteorology as a science and a profession. It is no secret that such operations are being promoted, under

17

such euphemisms as 'weather engineering' and 'weather management' projects, on the premises that the basic assumptions and techniques are already proven and that the remaining problems are largely of an engineering or logistic nature. But no smokescreen of management or decision-making jargon can hide the fact that the protagonists of this approach are continuing to use the same inadequate concepts and techniques that have failed to provide convincing answers during the past twenty years.[12]

But why are meteorologists so sceptical about the experimental results obtained so far? And why is it so difficult to arrive at definite conclusions? To understand this, we must look at the problems of scientific deduction and proof.

Physical deduction and statistical proof

The first seeding experiments had been conducted on stratus (layer) clouds. This is a type of cloud which is naturally stable and undergoes little or no change. If the same visible changes are observed after each seeding of stratus clouds, and if there is a plausible physical explanation for these changes, then their connection with cloud seeding is obvious and there is no need for detailed statistical proof. The same techniques are equally effective for the dissolution of ground fog, which, in principle, is just a layer of stratus in contact with the ground. However, for the purpose of increasing precipitation, this cloud type is not sufficiently productive.

The most important rain bringers are cumulus clouds, which are 'capricious' and unpredictable, as we have already learned. Their individual natural development cannot be anticipated with certainty, and because they do not always react in the same way to seeding, the efficacy of a modification technique cannot be easily assessed. To answer the question of whether the precipitation has really increased, we must resort to statistical comparison.

A primitive technique is the 'historical' comparison. Commercial rain makers often make statements to the effect of their cloud seeding have increased rainfall by 10% of the long-term mean in the area of operations. But such increases could equally have arisen from other factors of local or large scale. On the other hand, the maintenance of a certain mean rainfall amount against a natural downward trend could be a real achievement. For example, the 1973/74 famine in the Sahel zone could have been considerably

18

mitigated if the amounts of rain which fell on average during the long period before the drought (and were received again thereafter) could have been maintained.

For an objective assessment of scientific experimental results, observations on a 'control' object, not affected by the operation, are needed. For example, a chemico-biological experiment might be conducted on mice. Two groups of mice would be kept under identical environmental and nutritional conditions, one being fed with, the other without, addition of the drug to be tested. Under these conditions alone could one assume that effects observed only in the group of treated mice are the result of the drug and not coincidental. Similarly, during rainfall stimulation experiments, results obtained in the seeded area must be compared with precipitation in a control area the general weather conditions of which are similar, but in which no cloud seeding takes place. It is even better to compare changes from the long-term mean in the experimental area with the same changes in the control area.

There are, however, some difficulties. The natural range of 'random' variations – those which cannot be explained through present understanding of the problem – in the behaviour of different cumulus clouds is much larger than that which applies to laboratory mice which have been 'standardised' by selective breeding over many generations. Reference to 'general' weather conditions in the control area being similar to those in the seeding area already conveys the possibility that 'particular' differences may exist. These would be significant unless the test series were extended over many years. Also, it is not possible to achieve a strict separation between the atmosphere in the seeding area and that in the control area, in the way that the cages for experimental and control animals in the laboratory can be separated. It is quite possible that the effects of cloud seeding may be carried over the seeding area to the control area.

There are some statistical devices for increasing the significance of a test result.[13] The most important of these is the introduction of 'randomisation'. If the scientist in charge of the project is of the opinion that a situation exists which is particularly suitable for seeding, a straw poll (or any prearranged order not known to the leader) determines whether or not seeding is to take place. This method helps to find out whether the rainfall was really greater on seeding days with favourable weather than on nonseeding days with favourable weather. Also, seeding and control areas can be interchanged in random sequence, which eliminates a possible influence from differences in the weather characteristics of these areas.

In all these arrangements it is tacitly assumed that there are no long lasting after-effects of cloud seeding – though these could conceivably occur. Indeed, post-mortems on two American experimental series showed that seeding effects could be discerned 6 to 24 hours after the end of the operations, and that significant numbers of the dispersed icing nuclei remained in the atmosphere over these periods. If this fact was overlooked in the statistical analysis – i.e. if the period of after-effect was counted as a nonoperational control period – the test results became insignificant. On the other hand, if the after-effect periods were lumped with the operational periods, or cut out altogether, a real increase in precipitation amounts emerged.[14]

Finally, a 'moving target' can be used. This means that total precipitation from a seeded cloud is measured as it moves along and is compared with the total precipitation from a nonseeded cloud. In this case it must be taken into account that seeded clouds which grow into large cumulus towers with extensive ice caps may influence other clouds. Also, it is conceivable that a seeded cloud turns over more water vapour and produces more precipitation but, at the same time, detracts so much humidity input from other clouds that, in the final analysis, there is no net increase of precipitation over a given area.

This shows that each statistical method of cloud seeding assessment has potential faults and must be considered with a critical mind. Furthermore, it must be clear that statistical safeguards can make the occurrence of chance results less likely, but cannot exclude them altogether. Statistical results are therefore the more reliable, the larger the amount of experimental data. However, the number of situations under which cloud seeding in a given area is likely to be successful is limited and cannot be increased at will. If we do not wish to wait tens of years for results, we must draw statistical conclusions from fewer and smaller samples at the risk of a higher incidence of faulty deductions.

The 'pure' statistical proof takes only the final result into account. The primary assumption is that all members of the experimental population and of the control population are similar and comparable – otherwise, statistical treatment becomes impossible. But this assumption is not unconditionally valid in the realm of meteorology. There is a limit to the extent to which objects of the experiments can be compared. Finding out what really happens requires, besides a consideration of the statistical results, an examination of the way the basic data have come about.

An experimental series in South Dakota produced negative results. Analysis of the basic data showed that, by chance, less rain had fallen in the early morning before operations commenced on the randomly selected seeding days than on the control days.[15] This invalidates the statistical conclusion that there was a negative effect of cloud seeding; but it does not answer the question of what kind of effect, if any at all, had taken place.

Of course, the data on which the statistics are based are themselves not always reliable or representative. A raingauge measures precipitation on a small circular area, and rainfall amounts may be much higher or lower only a short distance away, especially during showers and thunderstorms. The measurements from a network of such gauges spread over an area give only an estimate of the total rainfall; an approximation which cannot be taken to reflect fully the small-scale differences which are produced by seeding. Some American scientists are of the opinion that at least in one experimental series the apparently negative results have probably arisen because there have not been enough raingauges in the target area to register the increase of precipitation actually produced by the seeding.[16] Radar measurements of precipitation are more suitable to assess the distribution of precipitation over an area, but the absolute accuracy of the derived total amounts is relatively low because radar picks up large drops better than small ones.

The analysis of data from many carefully conducted experiments apparently shows that in some cases positive and in others negative effects of cloud seeding occur, regardless of the final result being positive or negative. This is water on the mills of the sceptics who feel strengthened in their belief that cloud seeding has *no predictable* effect on precipitation and, to this extent, the doubts among

21

meteorologists are understandable. On the other hand, we can ask *why* the effects of seeding have been different on different occasions. In recent years, the viewpoint has been emerging that the main task of scientific work in this area is not the mere statistical analysis of the results, but an explanation of the processes which lead to these results. This idea is also emphasised in the WMO evaluation of weather modification, which says: 'It appears that the most sophisticated statistical procedures are an inadequate substitute for a more complete knowledge of the atmospheric mechanisms.'[17] This is true not only for precipitation enhancement but also for hail prevention.[18]

Thus proof for the success of a weather modification experiment rests on two pillars; statistical and physical. An experiment may be conducted entirely satisfactorily from the statistical point of view and yet not convince a meteorologist unless there is some plausible physical explanation for the effect that has been attributed to a certain action in a given situation. In other words, there must be reasoning for success as well as for failure.

One long-term experimental series in India, using a precipitation stimulation method especially suited for hot countries (cloud seeding with rock salt), produced a much greater success (precipitation increase by 40%) than could have been expected on theoretical grounds. Although the statistical procedures used in this experiment were impeccable, this result caused a lot of doubts.[19] It is, of course, conceivable that a satisfactory explanation will come foward in the future. But until then the question will remain open whether perhaps some other factors, not considered by the Indian weather makers, have produced the increases; or whether perhaps deficient measurement techniques, technological shortcomings or even manipulations may have simulated a better result than had actually been achieved.

The advocates of scientific 'precipitation management' must improve their theories and computer models until they can reliably predict how a given cloud will develop without modification, and how its development will change through cloud seeding. Such a prediction, which can be checked by observation and experiment, would have to cover not only precipitation, but also other statistically tractable data, like speed and extent of cloud growth, strength of updraft within a cloud, etc. The more agreement there is between

22

these predictions and reality, the better the theory, and the bigger the chance that concrete effects of seeding can be predicted correctly for individual cases.

Fortunately, the time is past when a modification experiment consisted of blind statistics applied to whatever raingauge data happened to be available. Hopefully, the time is passing when statisticians working in isolation, without knowledge of atmospheric processes, make pronouncements about weather modification. A new generation of statisticians is learning meteorology. Many are rolling up their sleeves to fly in aircraft, building on the knowledge gained of cloud behavior to design improved tests of hypotheses. Others are active participants in experiment design, using their statistical tools to specify gauging and radar accuracy requirements. Actual exposure to the atmosphere's complexities is a brutal, but effective, cure for the tendency to oversimplify or to daydream. Conversely, a growing population of meteorologists have recognized that they must learn and apply statistics.[20]

According to Dr Simpson such developments have contributed to an improvement in the ability to deal with more complicated problems, which requires meteorological experience as well as advanced statistical treatment.

What is a cloud?

A cloud is not a 'thing' but a 'process', an assemblage of water droplets and/or ice crystals floating in air and developing continuously. Development proceeds slower in stable stratus (layer) clouds than in the unstable cumulus clouds. However, in all cloud formations, there is a continuous interplay between forces and processes of building and those of dissolution.

A cloud is not a closed system. There is a steady 'metabolism' with the surrounding atmosphere. In particular, cumulus clouds draw in from below large masses of relatively moist air which they require for their maintenance. They also entrain relatively dry air through their sides which can lead to a 'dilution' and eventual dissolution of the cloud.

There are cases when a cloud builds up on one side and dissolves on the other, apparently moving in a direction different from that of the wind. A cloud cap on a mountain appears to remain motionless like a tuft of cotton wool, but the mountaineer, who moves inside it and perceives it as fog, can feel the air motion which is generally associated with it. Such an 'orographic' cloud produced by topo-

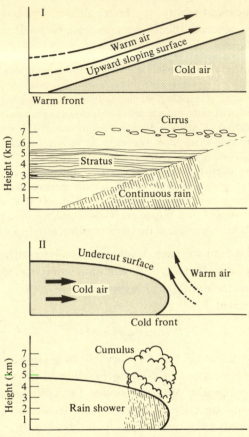

Fig. 4. Cloud formation by upslope motion of warm air above a ramp of cold air (I) and by a cold air thrust pushing the warm air upwards (II). (After K. Hammer and J. Neubert, *Lehrbuch der Physik*, Vienna 1959.)

graphy can best be compared with a waterfall; it remains at the same spot and maintains the same general shape, while water is streaming through it all the time. Similarly, on the mountain top, the water molecules move with the air stream through the cloud, which remains approximately in the same position.

The source material for cloud formation, water, is evaporated from oceans, lakes, plants and soil into the atmosphere. A given amount of air can take up a certain amount of water vapour only. This amount is the greater the higher the temperature of the air. If the temperature of air saturated with water vapour falls below the *dew*

24

point, some of the water vapour condenses as liquid water droplets (dew, fog, cloud). A good example for condensation by cooling in still air is the formation of night and morning fog near the ground.

Cooling will also result when air is forced to ascend to higher altitudes, as on a mountain in the afore-mentioned case of the cloud cap, or in forced upgliding of horizontally driven winds over a dome of cold air, or when a wedge of cold air undercuts a mass of warm air. Finally, heat can be transferred from the ground to the air above it, making the air buoyant and cause it to rise, thereby producing convection clouds.

Whatever the reason for the ascent, when the air rises it expands as a result of lower pressure and therefore cools. If it reaches a height at which its temperature falls below the dew point, water vapour may condense in the form of fog or clouds. This condensation into liquid droplets requires the presence of condensation nuclei – microscopically small particles in the air on which the condensation process is initiated. In absolutely clean air, completely free of dust and other impurities, supersaturation to the extent of several hundred per cent of relative humidity is required before condensation commences. Under natural conditions this problem does not arise. The free atmosphere usually contains an abundance of dust, salt and organic particles which act as condensation nuclei. The birth of a cloud is generally marked by the formation of tiny droplets with diameters from 2 to 40 micrometres (1 micrometre = 1 thousandth of a millimetre).

When water vapour condenses, heat is liberated – the same amount of heat which had been consumed in evaporating the water. Physicists talk of the 'latent' heat which is contained in water vapour. It warms the air around growing droplets, causing the air to rise further to even colder heights. There, the dew point is still lower – more water vapour condenses – more latent heat is given off – and so the cloud grows upward.

This development finally pushes the cloud top above the freezing level (zero degrees centigrade). The height of the freezing level in the atmosphere depends on geographical position, season, time of day and meteorological conditions. It is greatest in the tropics where it is found between 5000 and 6000 metres. It might be imagined that cloud droplets should freeze once they are cooled below 0 °C, but it has been known since the beginning of this century that this is not so. Balloon ascents and observations at the Sonnblick Observatory in the Austrian Alps have shown that clouds may consist of liquid water droplets, and not of ice crystals, at temperatures down to 10 and 20 degrees below freezing.[21]

The warming by condensation creates a strong updraft in the interior of the cloud the speed of which exceeds the fall velocity of the cloud droplets. If a cloud droplet happened to move below the base of the cloud, where the air is not saturated with water vapour, it would evaporate quickly. In order to fall through the strong updraft and to survive evaporation in passing hundreds or thousands of metres of unsaturated air, a drop of water must be much bigger than a cloud droplet Raindrops have a diameter of 1 to 6 mm (drizzle drops 0.2 mm). The mass of a raindrop is about a million times greater than the mass of the cloud droplets which form at the start of condensation. Such growth cannot be achieved solely by condensation on the cloud droplets, because abundant numbers of condensation nuclei compete for the water vapour. This brings up the question how rain forms at all.

One of the first to try answering this question was the German meteorologist Alfred Wegener, belatedly renowned for this theory of continental drift, which was not taken seriously during his lifetime. In a book published in 1911,[22] he pointed out that water vapour is attracted more strongly by ice than by supercooled water droplets of the same temperature. In the physicists' jargon: the saturation water vapour pressure over ice is lower than that over water. In a cloud, in which a mixture of supercooled droplets and ice crystals is present, the latter will grow at the expense of the former. This leads to the formation of larger and heavier solid 'hydrometeors' – snowflakes or hailstones, depending on conditions – which can no longer be supported by the updraft in the cloud, and will precipitate as snow or hail. Depending on the height of the cloud base above ground and temperature and humidity of the sub-cloud layer, this precipitation will either reach the ground unchanged, or will melt and arrive as rain. Sometimes, when the humidity below the cloud is very low, the precipitation may even evaporate before reaching the ground.

But how does ice crystal formation start in a supercooled cloud?

This question has occupied the Swedish meteorologist Tor Bergeron. In a publication of 1928, and later during the International Meteorological Congress 1935 in Lisbon, he put forward the argument that the air contains 'freezing nuclei' besides condensation nuclei. Freezing nuclei would, however, be less numerous than condensation nuclei, since the action of a freezing nucleus requires that its shape is compatible with the crystalline structure of ice. As the temperature decreases, the stringency of this requirement would be relaxed, since the 'crystallisation force', i.e. the tendency of water to freeze, would increase. Thus, he explained, the growth of a supercooled cloud to colder and colder heights would provide the required numbers of freezing nuclei. The German physicist W. Findeisen developed these thoughts further, made computations, and finally produced a proper theory of precipitation, based on his results.[23]

The temperature at which the number of ice crystals required for precipitation forms apparently depends on the kind and the number of freezing nuclei present in the air. Bergeron had already pointed out that the probability of icing is very low at temperatures above -10 °C. It is nowadays known that ice crystals will form only under special circumstances above -20 °C, and that liquid droplets may still be found at temperatures below -30 °C. At -40 °C, however, there is spontaneous freezing. Such low temperatures occur only at very great heights, in the tropics at about 11 km, and only few cumulonimbus clouds penetrate above this level.

Latent heat is liberated also during freezing, producing convective updrafts which carry the ice crystals. There are clouds which consist of ice particles in their upper parts and of droplets in the lower. There often exists a transition layer in which ice crystals grow at the expense of water droplets, then sink into lower parts of the cloud where there are more supercooled droplets, and there proceed to grow faster until precipitation occurs.

The critical temperature of -40 °C was discovered by Irving Langmuir and his collaborators during laboratory experiments to which we shall come back in the next chapter. Bergeron derived from this another idea about precipitation mechanism, the 'two-layer cloud system'. Ice crystals would form relatively easily in the colder upper regions, even without an abundance of suitable icing nuclei. Only a little snow falling from a cirrus layer or from the anvil of a cumulonimbus into the unstable stratification of a culumus cloud below it, he argued, could trigger copious precipitation easily.[24] Today, we know that this argument is right, and that precipitation of the two-cloud system type occurs relatively frequently.

The opinion of Findeisen, that precipitation can *only* form as a result of the process he described, has not been confirmed. American meteorologists have shown by theoretical considerations and observation that there is another precipitation mechanism: the coalescence of small droplets into larger drops by collision.[25] This requires, however, that there are relatively few and large (probably also hygroscopic) condensation nuclei in the air, which already on initial condensation produce comparatively big droplets. Such conditions are most likely to be fulfilled over the tropical oceans. In moderate latitudes this process produces, at most, small droplet drizzle from low stratus clouds. Real rain outside the tropics is practically always melted hail or snow.

2
Methods and applications

The hobby horse of a Nobel Prize winner

Irving Langmuir, the founder of weather modification on a scientific basis, received the 1932 Nobel Prize for Chemistry, in recognition of his investigations of adsorption and absorption on boundary layers and not for work in the field of atmospheric sciences. It was only towards the end of the Second World War that he embarked upon meteorology as an outsider. Without much regard for the opinion of specialists in this discipline, he repeatedly issued far-reaching and optimistic statements about the possibilities of planned weather modification. Since he was a highly respected scientist, these statements caused considerable attention and interest among the general public, in governmental bodies and in the scientific community. Meteorologists watched the activities of this intruder with growing unease. Being sensitive to frequent ridicule by a critical public, they feared that Langmuir's promises, which to them appeared premature and poorly founded, would damage the reputation of their discipline for a long time to come. They expressed their feelings in bitter words of criticism to which Langmuir replied in the same manner. Among meteorologists all this contributed to an almost instinctive and strong resistance to anything connected with weather modification.

Langmuir, born in 1881, studied under the German physicist and chemist Walter Nernst. From 1909 he was a member and later became director of the General Electric Company laboratories at Schenectady, New York State. He pioneered electric amplification and published trail-blazing papers on the theory of valency and other chemical problems. Also, he made a significant contribution to the development of gas-filled electric bulbs and to a welding technique employing atomic hydrogen. During the First World War he developed submarine detection equipment for the US Navy, during World War II he conducted research on smoke screens and on aircraft icing. This brought him into contact with cloud physics.

29

To gain better insight into the problems, the scientist – then over sixty – repeatedly visited, together with his collaborator Vincent Schaefer, the observatory on Mount Washington in New Hampshire.

Both scientists were passionate mountaineers and made the ascent to the observatory on foot several times during winter. At that time they had not yet read the meteorological literature and they were surprised to notice that clouds at the top of the mountain (1919 m) consisted of supercooled water droplets, even at temperatures far below 0 °C, which deposited rime on solid objects. This was nothing new, but being experimental scientists and used to an approach quite different from that of meteorologists, they did not leave it at that but decided to do some laboratory experiments in order to investigate the conditions which they had found to exist in the clouds.

Schaefer had started his career as a mechanic in General Electric's laboratories and had later been given the opportunity to transfer to scientific work. He came up with an experiment of genial simplicity. By breathing into a deep-freeze box, which had then just come on the market, he produced 'clouds': a perfect replication of natural conditions. The moisture in the breath condensed to fine mist droplets, but no ice crystals formed, although the temperature of the box was below −20 °C. During more than one hundred experiments using all sorts of icing nuclei Schaefer did not manage to induce the formation of ice crystals. He tried soot, volcanic ash, finely ground graphite, sulphur, silicates and other materials – all without success. The mist remained about 10 minutes and was then deposited as rime on the wall of the box.

On 12 July 1946, Schaefer was about to continue his experiments but could not achieve a low enough temperature in the box because it was a hot summer day. He intended to place a piece of dry ice (frozen carbon dioxide) into the box. The moment it made contact with the supercooled water droplets, thousands of small snowflakes formed. Even a tiny piece of dry ice was sufficient to produce this reaction. Schaefer soon realised that this was not because of the special properties of the frozen carbon dioxide but that the triggering factor was the low temperature. A pin dipped in liquid air produced the same effect. The critical temperature was around −40 °C. When moist air entered a cloud chamber which had been cooled below this temperature, ice crystals formed immediately, without any additives.

The crystals that formed during these experiments looked exactly like natural snowflakes.

At the time of these decisive experiments Langmuir happened to be away from Schenectady. When Schaefer told him of his successes he was enthusiastically excited. He immediately saw the important practical significance of these experiments: what Schaefer had achieved in the ice box could also be done in nature. Man had acquired a technique that made it possible to change clouds at will and to induce precipitation.

On 13 November 1946, Schaefer carried out the first test. From an aircraft he sprinkled three pounds of finely ground dry ice onto a supercooled stratus cloud deck. In less than five minutes snowflakes formed and fell about a thousand metres before they evaporated. A large hole opened up in the cloud sheet where the aircraft had flown. To be sure that this opening had not formed accidentally, more complicated tests followed. On one test run the trade mark of General Electric was cut out of a cloud bank. There could be no more doubt: seeding with dry ice caused snowfall and dissolved the clouds. The experiments made headlines and many papers quoted the optimistic remarks of Langmuir that applications of techniques of this kind would soon enable people to change weather at will. The Nobel Prize winner's new discovery stimulated interest among governments, public utilities, farmers who had been plagued by shortage of water and, of course, the military.

When the head of a big research laboratory has a hobby horse – and Langmuir's interest in weather modification was obviously his – it is likely to be infectious. Besides Schaefer, a number of other associates became concerned with the problem. One of them was Bernard Vonnegut, an icing nuclei expert, who had collaborated in the study on aircraft icing. He did not see much promise in Schaefer's random trials with various substances and held the opinion that one should search much more systematically. His primary concern was the shape which a nucleus particle should have in order to induce ice formation and he looked for substances with a suitable structure. Leafing through a manual on X-ray crystallography, he saw that silver iodide should have the required property, because the configuration of the atoms in a silver iodide crystal is very similar to that in an ice crystal. Vonnegut made an experiment in Schaefer's ice box when Schaefer had gone away for a few days. He was convinced it would be successful, yet – to his great surprise – the experiment was a complete failure.

On his return, Schaefer became interested in Vonnegut's ideas and himself started to experiment with iodoform and pure iodine. His results were not much better. A few weeks later, Vonnegut intended to try out the effect of metal-dust aerosols which he generated by sparks between electrodes. When he held a silver coin between the electrodes, the ice box was instantly filled with ice crystals. It happened just like in Schaefer's dry ice experiment, but the temperature was now well above -40 °C and the effect did not last. A repetition of the experiment on the following day produced no effect. Vonnegut was faced with a riddle.

The solution was soon found: traces of iodine from previous experiments were still present in the ice box. When the silver coin was introduced the first time, a chemical reaction produced silver iodide, the very substance which had been singled out as eminently suitable for providing icing nuclei. But this only raised another riddle: why did the first experiment with silver iodide not succeed? This time the answer proved to be surprisingly simple and, at the same time, very instructive to the experimental scientist: the 'silver iodide' which had been taken from the shelves of the laboratory was not pure. It had been contaminated and its surface had lost the typical molecular structure required to induce the formation of ice crystals. Vonnegut had blindly trusted the word on the label and had not even thought of checking on the contents after the failure of his experiment. Only a chain of fortunate accidental events eventually enabled him to prove his original arguments correct. Soon it was established that silver iodide of sufficient purity produces snowflakes from supercooled water droplets already at a temperature of -4 °C.

Vonnegut said later that purely scientific interest prompted him to perform these original experiments and that he did not then think of the practical implications of weather modification. But Langmuir's optimistic statements had triggered feelings of expectation and confidence. Vonnegut's discovery had opened a second way of seeding clouds – and both of them were at a cost which could be afforded by weather modification projects. Vonnegut also became tied up with further developments in this field. During the years that followed, he took a leading part in the development of burners and apparatus for the generation of silver iodide smoke for the purpose of weather modification, and participated in experiments which assessed their efficacy.

Naturally, after Schaefer's first successes, Langmuir also wanted to continue with the experiments. He realised that this could not be done indefinitely at the expense of General Electric, whose business interests were in a different field. Being director of one of the leading research institutes in the USA, however, he had appropriate connections and knew where to find the money. First, he succeeded in interesting the Army Signal Corps in his ideas; already in February 1947 a contract was signed according to which the field trials were going to be undertaken by the military and the laboratory work (against payment) by the General Electric Company. Later, the Office of Naval Research joined the project and the Air Force supplied aircraft and operating personnel. Finally, the US Weather Bureau joined in an advisory capacity. Thus 'Project Cirrus' was created, the first large scientific experimental project in the field of weather modification. The interest in this field of research which was aroused in American military establishments by this first project has remained to the present day.

Meanwhile, private enterprise also became interested in this strongly publicised new discovery. Farmers and agricultural corporations which operated crop-spraying aircraft began to inject silver iodide into the air, here and there, without knowing even the basic meteorological facts. Naturally, this attracted swindlers and speculators who saw a way of making quick and easy profit in this new activity. But these first years also brought the foundation of commercial weather modification companies by people who doubtlessly had a sound scientific education. One of the first large-scale commercial projects was run in California during 1947. Irvine Crick, head of the meteorological division of the California Institute of Technology (the famous Caltech), who supervised the operations in a scientific capacity, was so impressed by the idea of weather modification that he resigned his post and founded his own company of weather modification consultants. The city of New York had attempted to improve its drinking water supply by cloud seeding. At Langmuir's recommendation, this work was entrusted to the scientist Wallace Howell, who had collaborated with the Nobel Prize winner in studies of airframe icing on Mount Washington. Cloud seeding quickly became high fashion. In the early fifties, the area affected by operations of commercial weather makers in the USA extended to nearly 1 million square kilometres – about one-tenth of the area of the USA – and the turnover climbed to $5 million a year. But, when it gradually transpired that exaggerated hopes had been put into these techniques, the monetary turnover and the treated area soon decreased and finally stabilised during the sixties at $1–2 million a year, and 3–4% of the area of the USA, respectively. Langmuir himself continued seeding experiments on stratus clouds in New Mexico. In 1948 and 1949 he visited Honduras to familiarise himself with the techniques of a commercial weather maker who acted on behalf of the United Fruit Company. Earlier, in October 1947, he had tried to influence a tropical hurricane 500 km off the coast of Florida by cloud seeding, and when the tropical storm suddenly changed its track six hours after the operation, he immediately saw a causal connection between these events. However, meteorologists were able to point out that similar changes in tracks of hurricanes had occurred in some earlier storms, without human interference.

Nobody actually disputed that cloud seeding can cause clearly visible changes in clouds; many meteorologists however doubted that this could produce a significant increase in precipitation. To substantiate his statements, Langmuir developed his own statistical methods, but they were deemed to be inadequate by statistical experts. Throughout the last ten years of his life, the great scientist devoted his energy almost exclusively to the problems of weather modification. On his own initiative, he rediscovered a number of facts which had already been discovered by meteorologists but he also made some genuine discoveries. His involvement with this matter was characterised by tremendous emotion and much publicity. With hindsight, we could say that he was wrong on many points of detail and that he made claims which he could not substantiate. At the same time, there can today be no more doubt that two of the basic theses, which he propagated with the ardour of a missionary, are correct in principle:

1. that man is basically able to influence weather; and
2. that the main problem is finding the right triggering mechanisms

which can produce large consequences of a relatively small effort – the famous match which can set a whole forest on fire. 'This writer has heard colleagues say that the overly enthusiatic advocacy impeded the orderly progress of the science', writes H. R. Byers, professor for Meteorology at Texas University, 'but it probably can also be said that, without this pushing, governments would not have put their resources behind cloud physics research the way they did.'[1]

A prerequisite for the sympathetic attitude of the governments was the fact that Langmuir had managed right from the beginning to find an influential ally in the military. Matthew Holden Jr, professor of politics at the University of Wisconsin explains that the recognition of rain making as a matter which should be taken seriously came, like those of many other new developments, from proponents outside the relevant field (meteorology).

It was the intervention of a skilled and energetic outsider (Langmuir) that provided the nexus for military interest. Without that interest, the support for research could not have been achieved, which would have led to the self-confirming judgment that the idea of cloud seeding was a poor idea, in favor of which there was no evidence.[2]

A forecast which received no attention

While most meteorologists met Langmuir's inventions either with reservations or opposition, Tor Bergeron followed the idea of artificial weather modification with great interest from the beginning. In 1949 he published an article in the first number of the newly founded Swedish geophysical journal *Tellus* which dealt with the probability of increasing precipitation by cloud seeding.[3] In this article he noted that 'much publicity, and perhaps too much optimism has been bestowed on this activity.' Then Bergeron comes to the core of the problem: Any discussion which moves between the two extremes 'no artificial release proved' and 'world-wide rain control possible' misses the real point of the problem. One must rather examine 'whether such a release can produce an *appreciable* amount of rainfall, and *when and where* this could be done'.

After a quarter of a century of rather unpleasant quarrelling over the possibility of influencing weather, we are perhaps today in a better

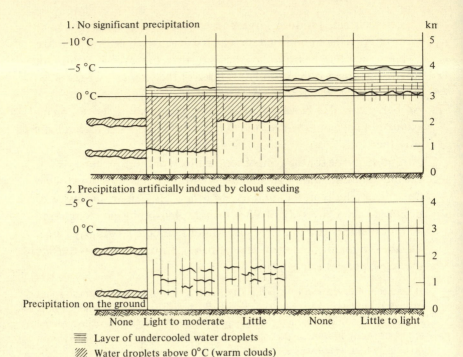

1. No significant precipitation

2. Precipitation artificially induced by cloud seeding

Precipitation on the ground

None   Light to moderate   Little   None   Little to light

≡ Layer of undercooled water droplets

▨ Water droplets above 0°C (warm clouds)

Fig. 5. Formation of precipitation from different types of stratified clouds according to Bergeron. *Above*: under natural conditions. *Below*: after cloud seeding. In the first column, the cloud layers are below the 0 °C level and seeding (with dry ice or silver iodide) is ineffective. In the other cases undercooled droplets do exist; therefore seeding can thin out or dissolve a cloud and cause a little precipitation, but this may evaporate partly or completely before it reaches the ground. (After an illustration by T. Bergeron in *Tellus* **1** (1949).)

position to appreciate the deep insight with which Bergeron looked at this question. Many senseless discussions and many inconclusive experiments could have been avoided if optimistic rain makers as well as sceptical meteorologists would have paid more attention to this remarkable piece of work.

Bergeron's basic thesis is that the seeding of clouds can only be successful in those situations where there is an adequate *continual* supply of supercooled droplets and, at the same time, an insufficient amount of ice crystals which trigger precipitation. That is to say, in situations where the available water in the cloud is not fully utilised. He pointed out, that this can occur only within a certain range of temperature which, as we now know, he estimated a little too high.

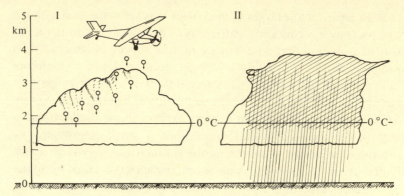

Fig. 6. Bergeron's idea of a cumulus cloud (I) being transformed into a copiously precipitating cumulonimbus cloud (II) by seeding. The aircraft need not fly above the cloud but may release the seeding material below the cloud base as the cloud sucks up air from below like a vacuum cleaner. (After an illustration by T. Bergeron in *Tellus* **1** (1949).)

In clouds which are cold enough for adequate numbers of ice crystals to form in a natural way, the available water is fully utilised. In this case, cloud seeding can lead to only an excessive supply of ice crystals ('overseeding') whereby very small droplets or crystals form, which either do not fall to the ground or are carried away by the winds to distant places. In clouds so warm that they do not contain any, or only very few supercooled droplets, seeding with ice crystal forming substances is obviously senseless. At temperatures that are only a few degrees below zero, the efficiency of these substances is severely restricted as there is not sufficient time for ice crystals to grow to the size required for the formation of precipitation. Between these two extremes, there should be an optimal temperature range for targeted modification – an idea that has been fully confirmed.

Bergeron then examines different cloud formations with respect to suitability for seeding and again comes to remarkable conclusions. For stratus clouds he predicts correctly that, at best, the rather limited amount of water already present in condensed form in the super-cooled droplets can be precipitated. As there is no rapid replacement of water, the result would probably be some light drizzle, or the precipitation might even evaporate altogether before it reaches the ground. The opening of breaches in the cloud deck could be useful for aviation, but no significant increases in rainfall on the ground should be expected.

37

With regard to the large cloud systems that are associated with the low pressure systems and their fronts in middle latitudes, Bergeron is of the opinion that an adequate natural supply of ice crystals is probably ensured. Seeding of these cloud systems is therefore unlikely to be either effective or profitable. Only a few investigations have been carried out so far on the seedability of frontal cloud masses. Most present-day American experts express the tentative opinion that in the case of frontal depressions a small re-grouping of precipitation intensity, and no significant increases over large areas, is all that can be achieved.[4] A more optimistic view is taken by Soviet rain makers, especially by B. N. Leskov, who thinks that the regular seeding of large cloud systems over the Ukraine could lead to precipitation increases of between 20 and 30% in winter.[5]

Regarding cumulus clouds, Bergeron thinks that seeding would be particularly effective when the heat liberated by the formation of ice crystals boosts the upward motion inside the cloud and pushes the vertical growth to the very high and cold layers where larger numbers of ice crystals can form in the natural way. Such precipitation increases would then result from an *indirect* effect of cloud seeding. This idea is a forerunner to what is called 'dynamic' cloud seeding by Joanne Simpson and others, and was tried out with success many years later. On the whole, Bergeron assesses the seedability of cumulus cloud rather sceptically, as he is of the opinion, that most of them would naturally precipitate near to their limit. His prognosis on this point is probably incorrect. Although a reliable method of modification for this elusive cloud type has not yet been developed, on the basis of theoretical insight and practical experience it can be stated with confidence that further efforts in this direction are likely to be successful.

Bergeron predicts good chances for stimulating rainfall in orographic cloud systems (produced by mountain ranges) within suitable temperature ranges. These are systems in which considerable amounts of supercooled water are produced for hours and days; they are not utilised in many cases, because clouds do not grow to the heights at which ice forms naturally. But for seaward-facing slopes of coastal mountain chains, an increase in precipitation may, as a rule, be undesirable, as precipitation there is generally abundant. Moreover, cloud seeding might produce a fall wind (foehn) effect in the interior

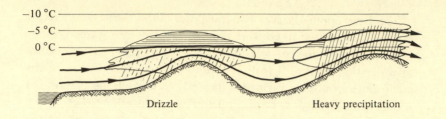

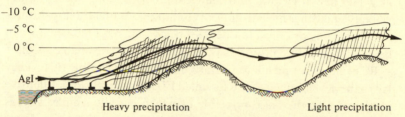

Fig. 7. Bergeron's idea of precipitation forming in orographic clouds naturally (above) and after seeding with silver iodide (AgI) from ground generators (below). The assumption that ice crystals would form naturally at temperatures between −5 and −10 °C is not correct; natural formation of precipitation requires considerably lower temperatures. (After an illustration by T. Bergeron in Tellus **3** (1949).)

land areas with detrimental effects on human well being. Alternatively precipitation on a seaward-facing mountain region might be reduced by 'over-seeding', while, at the same time, precipitation in the interior might be increased. This would in many cases be a desirable result.

All these forecasts have proved to be correct and the biggest seeding successes have indeed been with this type of cloud, as Bergeron had predicted a quarter of a century ago.

Fog dispersal – by brute force and otherwise

The only kind of weather modification which had already been developed before the discoveries of Langmuir and Schaeffer is the dissolution of fog at airfields. At the beginning, the technique applied was essentially one of brute force. The basic idea was that fog forms when moist air is cooled below dew point; hence, fog can be dissipated by warming the air to the temperature above the dew point.

The heat applied serves two purposes. First, large amounts of energy are required to convert the fog droplets into water vapour. (About six times the energy is required to evaporate a given amount of water than to heat it from 0 to 100 °C.) Secondly, the air of the environment must be warmed sufficiently to take up the evaporated water, without allowing saturation and over-saturation to occur. The amount of heat required varies from case to case. It depends on the density of the fog and the temperature of the air; also on the rate at which new fog flows into the cleared space, controlled by wind velocity and other meteorological conditions. With air temperatures above freezing or only a few degrees below, the required heating can be produced with conventional techniques at a reasonable price.

Relevant experiments were started in England during 1936. They led to the development of the so-called FIDO-systems (Fog, Intensive Dispersal Of), the basic principle of which is quite simple. Fuel is sprayed from perforated pipes along the runway and ignited. There are difficulties with occasional ignition failures and development of smoke but the system is on the whole successful. During the Second World War it had been installed at 15 airfields of the Royal Air Force. The first landing assisted by FIDO was made in November 1943. By the end of the war, more than 2500 aircraft carrying over 10 000 men had landed with the help of these installations; on the one occasion 85 aircraft succeeded within 8 hours.[6]

After the war, the American Navy tried to improve the system. In 1949, FIDO was installed at the civil airport of Los Angeles and remained in operation for approximately five years. The cost of the installation was $1 325 000. In spite of various technical improvements, problems continued with ignition, smoking burners and inadvertent issue of fuel. Moreover the system proved to be inadequate during dense fog. According to the view of the airport authorities an installation of adequate power would have been too expensive. Also, they thought then that this method would be made obsolete by the development of automatic landing systems. FIDO ceased operating at Los Angeles at the end of 1953.

The American Air Force continued to grapple with the problem of fog dissolution and concentrated on flexible techniques which could be applied without fixed installations. There were the seeding methods based on Langmuir's discoveries which could be applied to

40

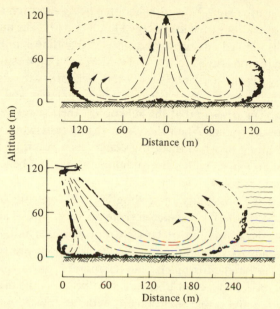

Fig. 8. Dissolution of 'warm' fog by a hovering (I) or slowly moving (II) helicopter. (After B. A. Silverman and A. I. Weinstein in Hess (a).)

the dissolution of a 'cold' fog which consists of supercooled cloud droplets. But, since fog in middle latitudes frequently occurs at temperatures above freezing – when ice crystal formation is not possible – other possibilities had to be explored. In recent years, the Air Weather Service of the American Air Force looked into various techniques but, according to a report published in 1974, had found no entirely satisfactory solution.[7] The main difficulty in dissolving 'warm' fog proved to be an insufficient guard against the flow of fresh fog by crosswinds or turbulence into the cleared area (fog is quite frequently associated with appreciable air movement).[8]

'Warm' fog is a rather stable weather condition in which one cannot apply Langmuir's 'match'. All methods of dispersal require a high amount of energy. Thin fog layers up to a height of a few tens of metres, which often form by nocturnal cooling, can be mixed with the overlying warmer air. This can be done by helicopters, whose rotors produce a strong downward draught. Thus fog can be temporarily cleared from small areas. In still air, the clearance will persist for about 5 to 10 minutes before the surrounding fog moves in and covers it up again. Thus, a helicopter being used for rescue or emergency operations can clear its own landing path. The method was successfully applied on several occasions in frontline supply and casualty evacuation missions in the Vietnam War.[9] Fog dissolution on particularly sensitive points of a road network can also be achieved in this manner. But, when applied to large and busy airports, a problem arises in that the helicopters themselves become obstructions to air traffic. When there is a strong crosswind which drives the fog along or when thick fog occurs as a result of large-scale mixing of air masses with different temperatures, the helicopter technique cannot succeed.

There is also the possibility of seeding 'warm' fog with hygroscopic (water-attracting) substances, such as rock salt. The salt particles cause the formation of larger droplets which fall to the ground. This leads to a diminution of the water vapour pressure in the air and the smaller droplets can evaporate more quickly. However, this process does not liberate latent heat. As the droplets in 'warm' fog are in a stable environment without upward motion, it does not come to the explosive multiplication of ice crystals which occurs in supercooled clouds. Therefore, a relatively large amount of seeding material is required and the particles must have a specific size which depends on meteorological conditions. If the hygroscopic particles are too large, they fall to the ground too fast and cannot take up sufficient moisture. On the other hand, if they are too small, they do not reach raindrop size and the seeding remains ineffective.

Many hygroscopic substances do not qualify for a practical application because they are damaging to the environment, or produce excessive corrosion on aircraft and airport installations and damage the clothing of personnel and passengers. These problems are familiar from the salting of roads. Other substances, such as urea, cannot be processed in the required particle size – they immediately disintegrate to very fine dust. Furthermore, the cost of seeding with such material is excessive. These problems are some of the reasons why, as yet, there is no practically useful seeding method for 'warm' fog.

With the advent of jet aircraft came the possibility of warming the air by the hot exhaust gases from jet engines thereby dissolving the fog. Initial studies of the application of such a technique were made

42

by the American Air Force during the operations of the Berlin Air Lift. The first provisional measures on military and civil airfields were simply the positioning of jet aircraft along the runway. As the engines were revved to full power, it was difficult to hold the aircraft in place. During a fog-clearing operation by a Caravelle aircraft at Zurich, a short-circuit occurred in the electric cables of the brake system and the airliner burst into flames. Moreover jet aircraft are not always lying idle when they might be required for such operations. For this reason one has reverted to fixed fog dissolution installations at airfields frequently affected by fog. The initial costs are high, but such installations are economic in the long run since they avoid diversions of modern large aircraft which have become much more expensive compared with the earlier small planes. Practical experience has shown that none of the automatic landing systems so far developed can provide the same safety as a 'visual' approach, where the pilot can see the runway.

After pioneering experiments had been carried out in France, an installation called 'Turboclair' was taken into experimental operation at Orly Airport near Paris, in 1970. It consists of 12 jet engines which are housed in concrete bunkers, 80–120 m apart, along the runway. The engines ignite automatically and are remote-controlled from the tower. The installation can keep a 1.4 km section of the runway free of fog, from 300 m beyond the threshold to 800 m beyond the touchdown reference point. The capital outlay for this installation was around $3 million; the fuel used under average meteorological conditions is about 1500 litres per landing. Turbulence caused by the rising exhaust gases does not present undue difficulties for landing aircraft. The noise of the installation is less than the noise made by the landing aircraft but is superimposed on the latter. There is no excessive air pollution from the combustion products.[10] Encouraged by the results of the experimental operation, the French Ministry of Transport commissioned the Turboclair system for operational use at the Paris airports of Orly and Charles de Gaulle in 1974. During the winter of 1975/76 this system enabled 126 landings, involving altogether more than 10 000 passengers during 78 hours of fog (combined at both airfields).[11] During the winter of 1976/77, 128 landings were made possible during 60 hours of fog. 'This fog dispersal system is thus found to give full satisfaction technically and has consequently required no significant technical development these last two years', writes E. Sauvalle of the Aeroport de Paris in a letter to the author.

Since 1970, the American Air Force has developed a similar system. Instead of jet engines it uses burners specially constructed for this purpose, with the aim of achieving relatively low combustion temperatures to reduce the risk of fuel explosions as much as possible.[12]

A number of other techniques, like fog dissolution by seeding with electrically charged particles, seeding with soot to increase uptake of heat, fog prevention by surface-active substances, fog dissolution by ultrasound, etc., never got beyond the stage of theoretical discussion or laboratory experiment. The heating of runways by underfloor pipe systems is conceivable in cases where a power station, producing cheap waste heat, exists near an airport, but this has not yet been tried anywhere. The capital outlay for such a system would be high, its running costs would be small. However, whether the heat supply would be sufficient to clear thick or drifting fog is another question.

While 'warm' fog is stable and can only be removed with 'brute force' the Langmuir–Shaefer technique can be employed for the dissolution of 'cold' fog. Once a number of ice crystals have formed, they grow at the expense of the supercooled water droplets until they fall to the ground as snowflakes. Latent heat is liberated during this process, which warms the air and evaporates other fog droplets. In this way a strong effect can be achieved with relatively small amounts of seeding material. In a single flight along a straight path, seeding with dry ice can clear the fog along a channel up to 3 km wide for 30–60 minutes.

'Cold' fog can also be seeded by ground generators. The liberation of latent heat produces turbulence which carries the icing nuclei to the upper limit of the fog.

Instead of using dry ice to produce the required low temperatures, one can also spray liquid gas (propane) which evaporates on leaving the pressurised tanks and takes up heat from the air (this is the principle on which refrigerators work). Liquid gas has the advantage that it can be stored in pressurised vessels at normal temperature, while dry ice must be stored at very low temperatures (below -80

°C). At Orly airport pioneering work has also been done with this technique. Already in 1964, a plant with 60 propane sprayers controlled from the tower was taken into operation. During the winter of 1970/71 this installation enabled 340 landings and 284 take-offs.[13]

The dissolution of 'cold' fog is the only technique of weather modification which has so far gone beyond the experimental stage, functions reliably and is in routine use at many airports. American civil aviation has saved around $900 000 at a cost of $80 000 with such operations during the winter of 1969/70. The US Air Force has installed propane sprayers at some military airfields in the USA, Alaska and Germany, which have proved their worth in flat terrain under steady wind conditions. However, where topographic features are close to the airfields and complicate wind conditions, they are of little use.[14] Dry ice is used in places that have no permanent installations. During the years 1968/74 the application of these techniques facilitated 2741 take-offs and landings of US military aircraft. Even during the relatively fog-free year 1974, about $25 000 could be saved.[15]

In the Soviet Union, fog and low stratus clouds are being dispersed by seeding with dry ice on at least 15 large civil airports.[16] There are no data available on the military uses in the USSR. Laboratory experiments have been made with a mixture of propane and silver iodide, the latter allegedly being much more effective than pure propane at temperatures above −4 °C. Nothing is known about operational applications of this technique.

The Research Institute for Air and Space Travel in Oberpfaffenhofen in the German Federal Republic has carried out trials with sprayers mounted on trucks which disperse liquid carbon dioxide into the fog. Preliminary trials have shown that liquid carbon dioxide, which is much less difficult to store, produces the same effect as dry ice. Compared with propane, liquid carbon dioxide has the advantage that it is not combustible and therefore less dangerous. Good results have been achieved at air temperatures of −1.5 °C and below. But the trials in the Oberpfaffenhofen airfield also show that one mobile unit is not sufficient. Under variable wind conditions, at least three units must be employed to keep the runway free of fog.[17]

To seed clouds with silver iodide, either silver iodide smoke can be generated at ground level or the seeding material can be carried by aircraft, rockets, or shells into or near to the cloud.

Ground generators burn a solution of acetone which contains silver iodide and the ascending gases and air currents should carry at least part of the freezing nuclei to where they are required. In flat country this hope is often dashed unless the cloud base is very low. There is a better chance of success in regions where a large-scale air current moves over a ridge of high ground and thereby ascends. If the ground generators are placed on the windward side of the mountain range, there is a fairly good chance that the silver iodide reaches its target.

Ground generators are less expensive than aircraft and they can be operated uninterruptedly for days. The operation of cloud seeding aircraft is far more restricted financially, fraught with operational hazards during cloudy weather, and practically precluded during night time over high ground.

Aircraft are mainly used for the seeding of cumulus clouds over flat ground. This type of cloud sucks in air from beneath like a vacuum cleaner and an aircraft need not actually penetrate the cloud. It can operate below the cloud base and release the silver iodide there – either by flame generators, similar to those used on the ground, or by pyrotechnic devices, so-called 'flares'. In these, the silver iodide is mixed with the fuel or oxidation substance.

High-altitude aircraft – which may be available when the military sponsors such operations – can fly above the clouds and drop the flares into them. This ensures that as many icing nuclei as possible get into a particular part of the cloud, as is necessary for hail prevention seeding.

Rockets are used mainly in hail prevention operations. The seeding material is either dispersed at the maximum height of the trajectory by an explosive or, as in the case of flares, added to the propellant. This allows a more extensive seeding along the line of trajectory. Small 'hail rockets' with an altitude range of 2000 m can be bought in western and central Europe by farmers and used by them as they see fit. Their effectiveness is doubted by specialists, particularly since the cloud base is rather high in middle latitudes during summer and only a small proportion of the seeding material gets into the cloud. Also, these rockets carry only 15–20 g silver iodide, which is far too little to seed a thunder cloud effectively.[18]

Considerably larger rockets, which can carry several kilograms of seeding material, with considerable accuracy, to altitudes of 6000–14000 m and shells charged with silver iodide are used in the Soviet hail prevention operations. Thin cartridge walls and a special construction reduce the risk of large fragments falling to the ground after the explosion.[19] The use of such massive cloud seeding carriers requires specially trained (artillery) personnel and can only be done in areas with relatively low air traffic density. Before the start of a hail prevention or cloud modification operation, contact must be made with the appropriate air navigation authorities and air traffic control must ensure that no aircraft will enter the operational area. If this is not possible, operations have to be delayed until the air space is clear. The cost of re-routing aircraft because of rocket operations must be measured against the expected benefit from hail prevention operations. It is conceivable, that large rockets containing seeding material could be carried by aircraft to directly below the target clouds and be fired from there. The expense of such a seeding operation would be greater but air traffic would presumably be less impaired; also, the operational range of hail prevention operations would thereby be enlarged.

Conventional ground generators or flares do not allow control of the size and surface properties of the liberated silver iodide particles. Laboratory experiments have shown that the efficacy differs with the size of the icing nuclei. Depending on cloud temperature and other physical conditions (saturation or supersaturation with water vapour), silver iodide particles of different sizes should be available, if the maximum effect is to be obtained from a given amount of seeding material. For this reason, US technicians have developed a special burner which allows the production of silver iodide crystals of different size and characteristics, according to requirements.[20]

Rainfall stimulation – posing the right questions

In 1961, Jerzy Neyman and Elizabeth Scott, two respected statisticians of the University of California at Berkeley, published a survey showing the results of rainfall stimulation activities. Their conclusions were shattering. Of all experimental series which had been carried out until then, only 23 had been based on proper statistical planning and had lasted long enough to provide significant results. In only 6 of the 23 experiments, cloud seeding could be proved successful in having increased the precipitation. In 7 cases, the results were not statistically significant and in 10 cases, precipitation not only was not augmented but was apparently considerably reduced.[21]

Had the doubts of the sceptics been justified? Had the rain makers pursued a *fata morgana*? At first sight it appeared impossible to draw another conclusion. However at closer inspection these conclusions turned out to be a typical case of what Joanne Simpson calls 'blind application of statistics'.

There is agreement on Neyman and Scott having judged correctly from the viewpoint of their profession. Many of the weather modification projects were indeed planned in a statistically unsatisfactory way. There is no doubt about the commercial weather modifiers having applied a technique which had not yet been proved scientifically. Since they had been paid by their clients to deliver rain rather than to advance scientific knowledge, they had given little or no attention to the possibility of evaluating the results of their operations statistically. Those who had tried to back up the successes of their operations by statistics were criticised by statisticians for the way they handled their data. Also a number of meteorological experts, who are not experts in statistics, had underestimated the difficulties associated with evaluating experimental results. Their experiments were badly planned and inconclusive – especially if they wanted an answer to the general question of whether precipitation can be increased by cloud seeding.

But even more important than these shortcomings is that the general problem was itself wrongly defined and no meaningful answers could have been produced, even with faultless statistical planning. It is tempting to assume that many of the meteorologists who conducted experiments on artificial stimulation of rain during the 1950s and '60s had never read Bergeron's fundamental paper of 1949. Otherwise they would have realised earlier that they should not have posed the question of '*if* cloud seeding is promising at all' but rather '*under which circumstances*'.

'Cloud seeding started in Australia in 1947', cloud physicist J. Warner told the Tashkent WMO Conference in 1973. 'After preliminary experiments in seeding of single cumulus clouds, a series of area experiments was initiated, the first in the Snowy Mountains, between 1955 and 1959. For its day, the statistical design of this experiment was good. In later experiments in Southern Australia, New England and Warragamba a more powerful statistical method was employed'. Yet, the overall results of these four experiments are, according to Warner, 'unconvincing and do not justify any definite statements as to whether seeding had increased the rain or not.[22]

Australian cloud physicists had pointed out as early as 1949 that seeding success was closely related to cloud-top temperatures.[23] Nevertheless, apparently not enough attention was paid to this important point. With regard to the Australian experiments, Warner explained in his summary at Tashkent:

In spite of what appeared to be good statistical design, the degree of attention paid to the meteorology of the experiment did not allow many desirable analyses to be made. It seemed that the effects of seeding were more complex than expected and there were strong indications in the results that while seeding had increased rainfall on some occasions, on others it had reduced it. In retrospect, it is clear that identical observations (of cloud depth, cloud top height and temperature, etc.) should have been made in both areas during unseeded as well as seeded periods throughout each experiments.[24]

Contradicting results arose not only in Australia. Yet wherever not just the final precipitation was measured but other meteorological processes too, it could be shown that these contradictions do not arise accidentally but follow a definite scheme. The Israelis, within the framework of their second experimental series from 1969 began to investigate the relevant physical factors. Their most important conclusion was that seeding is successful only when temperatures at the upper limit of the cloud are between -13 and $-25\,°C$.[25]

The 'blind' statistical investigation of rainfall data in the American project 'Whitetop' had, according to Neyman and Scott, shown a negative final result.[26] Changnon showing deeper meteorological insight, later contended that the apparently negative outcome was probably a result of an insufficient number of rainguages having been installed in the target area.[27] However, as the final result of this experiment is accidental it is not really relevant. What is decisive is

49

the result of the data analysis, which shows that seeding produced a marked *reduction* in precipitation when the upper limit of the cloud extended beyond 12200 m altitude, and a marked *increase* when the upper limit of the cloud was between 6100 and 12200 m. No significant changes were produced when the upper cloud limits lay below 6100 m.[28] Since there is a close relationship between the altitude of the upper cloud limit and its temperature, these results prove that there is an optimal temperature range for cloud seeding – as Bergeron had predicted in his hardly noticed paper of 1949.

Of course such numerical results can not be transferred from one place to another. 'In our situation we must be careful to seed clouds whose tops are cold enough', writes the Australian weather modification expert E. J. Smith in a letter to the author of this book, 'whereas in Colorado the reverse is true'. Detailed investigations based on general theoretical knowledge must be made in each region in which cloud seeding operations are planned, to check how these can be applied under specific local conditions. Recently, such investigations have shown that there are other factors besides temperature which determine the chances of successful cloud seeding, such as windspeed, and the velocity of cloud movement.[29]

It is the aim of these investigations to come up with a 'recipe' which outlines concrete meteorological conditions promising for effective seeding in a given place, and ways of performing such seeding so that the desired effect may be achieved. Operations that are not based on such fundamental considerations are 'blind' and do not allow the prediction of results with confidence. Of course, we could imagine that a commercial weather modifier who has worked for many years in the same area intuitively assesses these conditions on the basis of his personal experience and is thus capable of a basically correct decision.

Water from the cloud cap

Of all the US states, California has had the greatest increase in population during recent years. High immigration, a natural growth in population and rising standards of living have produced a strong increase in demand for water. Although some precipitation occurs on the western slopes of the mountains, the local reserves of water are insufficient. Los Angeles has to import most of its drinking water

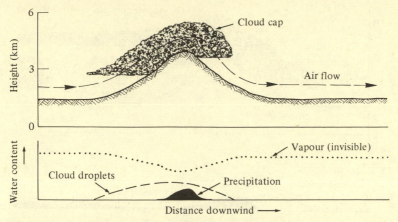

Fig. 9. Schematic formation of orographic cloud caps. When there is no precipitation, the cloud base is at the same height on either side of the mountain. The lower strip shows the amounts of invisible water vapour (moisture), cloud droplets and precipitation. (After R. D. Elliot in Hess (a).)

by a 500-mile pipeline from the Colorado river. Hydroelectric power stations and the irrigation installations in the San Joaquin Valley suffer from water shortage. Tor Bergeron's contention, that mountain slopes facing the coast usually receive sufficient precipitation, does not apply in this case. There have been projects in California and other states in the western USA concerned with the augmentation of rainfall by cloud seeding since the end of the 1940s.

The 'theoretical' background developed by commercial weather modifiers was rather fragmentary at the beginning, but it contained some basically correct arguments. Initially there had sometimes been seeding of clouds which were probably generally too cold for significant effects to take place; but conceivably the activity of the rain makers contributed to the winter snowfalls and thus, indirectly, to a better water supply in the summer. It appears that the Pacific Gas and Electric Company, the Los Angeles Flood Control, and other customers did not spend their money completely in vain.

The orographic cloud systems which were seeded in Western parts of the USA generally formed before and after the passage of frontal systems. During the passage itself, the meteorological conditions are complicated and operations are difficult to plan. However during periods before and after the passage, conditions are stationary and suitable for seeding. Cloud caps form on the mountain ridges and

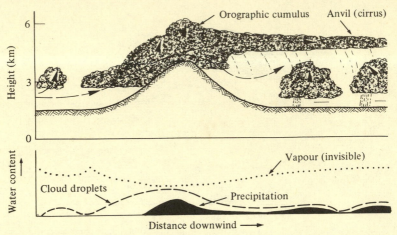

Fig. 10. Schematic formation of orographic cumulus clouds on passage of a front in the mountains. (After R. D. Elliot in Hess (*a*).)

often remain for hours, or even days, sometimes extending over hundreds of kilometres along the mountain chains. Although Los Angeles has the same geographical latitude as Algiers, temperatures in the hinterland mountains are sufficiently low in winter for the cloud caps to cool below freezing. Yet often they are not cold enough for the natural formation of ice crystals which trigger precipitation. Seeding is normally done by ground generators, since aircraft operation in the mountain areas is too problematic. Consequently, the seeding material does not always reach the best place. 'The delivery of seeding material probably remains the weekest link in operational programmes' says Professor Lewis O. Grant of Colorado State University, who has for many years been involved with the artificial modification of orographic cloud systems.[30]

In mountain valleys, the cold air often builds up underneath a layer of warm air and forms so-called temperature inversions above it. These prevent the silver-iodide-containing smoke from ascending. To get as close as possible to the clouds, the generators have been shifted higher and higher into the mountains. The head of a large Californian firm of weather consultants, R. D. Elliot, describes how he tried to make arrangements with people owning cabins in the mountains to use them while they stand empty in winter; he operates his installations from them and looks for young couples who are prepared to take over the routine work of such stations.

During the 6 or 8 months that they were snowed in, communication with them was carried on by radio. Surprisingly it was not difficult to find couples willing to do this work since many people find this type of isolated life a challenge. In spite of their enthusiasm various sorts of problems arose as a result of the deep snow accumulations. One man skied out during winter, leaving his wife forever. Fortunately we were able to bring a replacement couple in, and her out, in one snowcat trip. A couple ran out of fuel for the radio motor generator – they had used too much to run their hi-fi. Another man put his extra time to good use. He wrote his doctoral dissertation on cotton culture.[31]

Intent on avoiding the human problems and on reducing the operational costs for these stations, remote-controlled automatic generators started to come into use during the late fifties. Some of them were located so remotely that the supply of fuel and seeding material, and the maintenance had to be carried out by helicopter.[32]

Commercial cloud seeders operating in Cuba met local problems of a special kind during the fifties; each time one of them set out to attend to his generators in the mountains, he had to get special permission of Batista's government – then still in Havana – and from Castro's guerillas which already controlled the mountain regions.[33]

The seeding of orographic cloud systems was, like the dissolution of 'cold' fog, another application of Langmuir's ideas which worked out in practice. In 1957, a committee established to advise the President of the USA, had already discovered, by statistical investigation of a large number of commercial projects, that an increase of snowfall of 10–15% could be achieved by this method and that these results were not just accidental.[34] Later, experimental series conducted under observation of strict statistical precautions, have confirmed this result in principle. There are good reasons to assume that in climatically favourable places even better results – precipitation increase up to 30% – could be achieved.[35]

By far the most thorough scientific experiments in this field was conducted by Professor Lewis O. Grant in the 1960s on a 3000 m mountain pass near Climax in Colorado, to the west of Denver. Theoretical considerations and practical experiments went side by side. Grant first studied the wind situation on a scale model in the wind tunnel, developed a mathematical computer model of the processes, and investigated the efficacy of seeding material under varying conditions in the cloud chamber. In his field experiments he used a whole arsenal of ground measurement apparatus, radiosondes, free-floating baloons equipped with sondes, paper and plastic kites. This allowed him to determine fairly accurately where the seeding material issuing from the ground generators moved, and what effects were triggered in the cloud. As expected, these trials again showed that cloud seeding works only when the temperature is within a certain range and other conditions are right. However, favourable conditions occurred so often that the expense of the operations was worthwhile. On the basis of his experimental results, Grant eventually developed a 'recipe' for cloud seeding operations on Climax pass. Practical rules of this kind cannot be schematically transferred to another area – but the method used to arrive at this knowledge can.[36]

The seeding of orographic clouds cannot be successful unless the seeding nuclei have sufficient time to induce the physical processes which lead to the formation of precipitation. In a Wyoming project there was no difficulty in increasing the concentration of ice crystals in the cap of the Elk Mountain. However, it did not produce an increase in snowfall because the mountain range is narrow and the winds blowing along it were relatively strong; the ice crystals produced by the seeding leave the cloud and evaporate before they have time to grow to the size at which they could precipitate as snowflakes.[37]

The apparent success of many operational seeding programmes in augmenting the water in reservoirs of hydroelectric stations, the success of the experiments of Grant and others in the USA, the Australian achievements in increasing rainfall on the watershed of a hydroelectric power station in Tasmania,[38] and the successful operations on orographic clouds in northern Israel induced the authorities in the United States to start a large-scale pilot project in the San Juan Mountains of Colorado in 1970. Thirty-three ground-based generators, 13 of them remote controlled, the others operated by local farmers and ranchers in telephone contact with a central control, are used for the seeding. The objective is to increase the snowpack in an area of 8500 square kilometres (the size of Wales) so that an additional 300 million cubic metres of much needed water would swell the Colorado River.[39]

By 1974 it became apparent, however, that this project was not likely to produce statistically significant results. According to a letter from Professor Grant to the author, this was mainly because of the following reasons:

1. There were problems in forecasting seedability criteria for 24 hours ahead. Seeding was carried out on many occasions when conditions were such that positive effects could not be expected. This would not have occurred if one had forecast conditions correctly.

2. The number of seeding operations was too small because of shut-downs resulting from excessive snowfall, avalanche danger, forecast problems etc.

3. Physical observations scheduled to be made before the programme started were either not made at all or started only later in the programme. Thus, the experiment was carried out partly 'blindly', without a sufficient scientific basis.[40]

While point 1 is very important in a *randomised* project, it is much less so in a fully operational programme, since the operator can adjust to conditions which are expected during the next several hours. An operational programme of this kind is now running for the third consecutive winter.

No decision has yet been taken on starting a very large seeding programme for the whole Colorado Basin which, according to its advocates, should bring an additional 2400 million cubic meters of water – which would mean an increase in streamflow of 20%. The cost of such a scheme is estimated at $10 million a year, but considerable diversity exists in the estimates for the potential benefits: pessimists assume that the project would just about pay for itself, optimists predict an 800% profit, possibly rising even higher as food prices increase. On top of this come the beneficial effects of reducing the water shortage which are not measurable in terms of dollars. In any case, all other presently known ways of bringing additional water, or of reducing its consumption, are much more expensive.[41] On the other hand, the disadvantages likely to arise in the target area of precipitation are relatively small. They are mainly increased costs for the removal of snow. At present, seeding operations are terminated

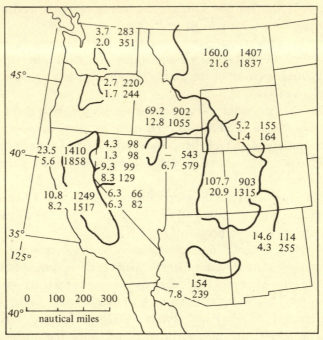

Fig. 11. Summary of average potential incremental streamflow. In each set of numbers left column gives area in thousands of square miles of the total basin (top entry) and massif area (bottom entry). Right column gives incremental runoff in thousands of acre feet for the conservative estimate (top entry) and the liberal estimate (bottom entry). (After R. D. Elliot and W. Shaffer, *Preprints 4th Conf. on Weather Modification, Fort Lauderdale, November 1974*, Am. Meteor. Soc., Boston, 1974.)

when the snowpack exceeds 150% of the long-term mean, when there is an imminent, or a predicted, danger of avalanches, when blizzards are expected, or when there are other serious reasons. Seeding on Climax Pass was once terminated to prevent further hardship to Christmas traffic on the roads.

An approximate calculation for the twelve most important river basins in the western USA shows that the river flow might be increased by at least 10 million cubic metres a year by cloud seeding.[42]

The idea, put forward by Langmuir and Bergeron, that 'over-seeding' can produce a large number of small ice crystals, which will rather be carried away than precipitated at the place of seeding, has been fully confirmed. The determination of the amount of seeding material required in each particular case is still problematic.

56

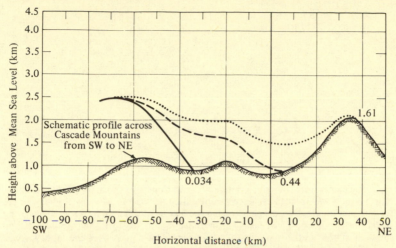

Fig. 12. Trajectories of different size snowflakes above the Cascade Mountains. Single (large) flakes (one per litre of air) follow the solid curve; the dashed curve represents the movement of smaller flakes (25 per litre), the dotted curve that of very small flakes (100 per litre). The longer the snowflakes remain airborne, the more water vapour is converted into snow. Thus the amount of snow per litre of air (figures along the base indicate milligrams of snow) increases with the distance over which the snowflakes are carried. Excessive seeding (overseeding) can produce a large number of small snowflakes. (After P. V. Hobbs and L. F. Radke, *Proceedings of the WHO/ IAMAP Scientific Conference on Weather Modification, Tashkent, 1973*; WMO 1974; p. 165.)

However, some experiments, especially a long-term experimental series in the Cascade Mountains, in the State of Washington on the Pacific coast of the USA, have shown that, under given conditions, snowfall can be shifted from the windward side to the leeward side of the mountain.[43]

Application of this method is not restricted to mountainous areas. It might also be used in the region of the North American Great Lakes. Their eastern and southern coasts are often affected by heavy blizzards when polar air moves over the relatively warmer water surface during the winter. It is hoped that by overseeding the snowfall could be redistributed over a larger area[44] (see fig. 13, p. 58).

If this technique can be developed to perfection in the future, it could be used for the prevention of flooding in certain situations. The aim would be to shift part of the precipitation from the catchment area of a flooding river across the watershed into a less endangered river basin.

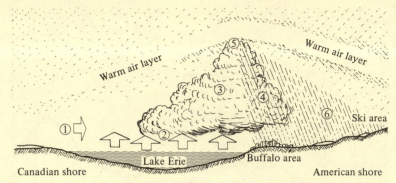

Fig. 13. Temperature inversions frequently lead to peculiar weather situations in the North American Great Lakes region during the cold season, whereby cold dry air (1) flows from Canada towards the lakes. On the mixing of this cold air with the relatively warm, moist air above the lakes (2), massive clouds (3) form and produce heavy snowfall at the southern shores (4). Overseeding of clouds over the lakes (5) can redistribute snowfall over larger areas (6) and possibly benefit the skiing resorts further south. (After W. R. D. Sewell (b).)

In search of a water generator

Cumulus clouds are of immense importance to man's livelihood and to his life. First they produce more than three-quarters of the rain that waters our planet. Second, giant cumuli constitute the firebox of all severe storms, such as the hurricane, thunderstorm, hailstorm, tornado, and squall line. Third, cumulus clouds are a vital part of the machinery driving the planetary wind systems; and fourth they act as a valve for regulating the income and outgo of radiation from the Earth

writes Joanne Simpson.[45]

According to her, the controlled modification of cumulus clouds and other convective cloud systems offers enormous potential benefit; drought mitigation, increase of food production, the possibility of producing more energy and of combatting the accumulation of toxic substances in the atmosphere. This would have far-reaching significance in view of the threatening shortages in food and energy supply. But, 'after a quarter century of effort there are only a half dozen or so examples in which a cumulus modification hypothesis has been conclusively demonstrated to work. However, in face of the difficulties involved, the fact that there is even *one* cumulus modification experiment that is conclusively successful demonstrates that cumulus clouds *can* be modified.'[46]

58

As we can see, Dr Simpson applies strict rules to the assessment of her own work. Of the successful experiment she not only expects an increase in the precipitation, but also an increase of scientific knowledge. For only when we know *how and why* success has been achieved is there a prospect for repetitive successful modifications.

Yet neither the farmers who are bothered by drought and hail nor the commercial weather modifiers are prepared to wait until experiments have solved scientific problems. They are like very ill persons who have heard of a new drug with tremendous healing power. They turn to scientists, politicians and civil servants demanding help. What stand can the scientist take in this case?

An attempt to answer this question was made by the meteorologist W. F. Hitschfeld of the University of Montreal during the Fourth American Conference on Weather Modification, at Fort Lauderdale, 1974. In his opinion the scientist must first make it clear that his medicine has not yet been proven; therefore each application must be coupled with a thorough observation and statistical analysis so that the value of the drug can be gradually established. A politician, who depends on the goodwill of the electorate, might not see this point so clearly. But he should also be aware that excessive application of techniques which have not yet been tried only increases the flood of data and blurs the signs which indicate whether or not hail damage has been reduced or precipitation increased.

The idea that 'seeding is cheap' and hence that a little success, measurable or not, pays for it, and in any case 'keeps the farmer happy' is poor politics and even worse medicine. A placebo may enhance a patient's powers of recuperation, but the clouds overhead respond neither to optimism nor pessimism. Moreover, if we cannot measure small improvements resulting from seeding, how can we be sure that seeding might not lead to increases in damage? It is as easy to construct qualitative reasons for this effect as for the opposite.[47]

Cumulus clouds are carriers of the most copious precipitation and have been the preferred guinea pigs of weather modifiers from the start. However, they are of elusive character and their mathematical modelling, as well as the assessment and theoretical interpretation of experimental results, is much more difficult than that of orographic clouds. 'I suspect that part of the problem is that with orographic situations, rain is relatively regular and uniform: if a change can be

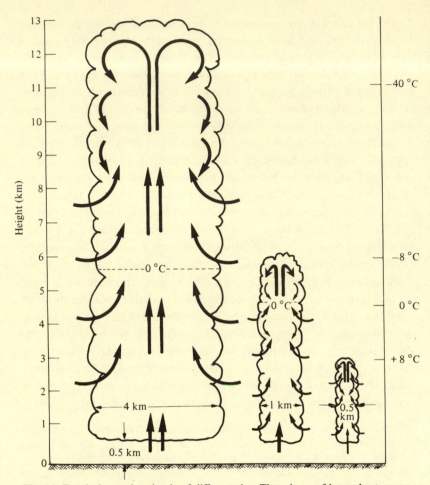

Fig. 14. Tropical cumulus clouds of different size. The release of latent heat causes the temperatures to rise inside the cloud. Thus the 0 °C level inside the cloud is above that in the surrounding air. The cloud on the extreme right does not even reach the 0 °C limit and does therefore not contain undercooled droplets; thus only seeding with rock salt or other hygroscopic substances might be effective. Precipitation could be triggered off by seeding with silver iodide in the upper part of the medium-sized cloud, while no precipitation is likely to form naturally. The giant cloud on the left has grown into upper regions where icing occurs naturally. (After J. Simpson and A. Dennis in Hess (a).)

caused it can be detected' writes the Australian cloud physicist E. J. Smith in the afore-mentioned communication with the author. 'In convective situations, rain is more scattered in space and time; therefore, even if a change in the rain can be caused, it is difficult to detect.'

Most cumuli dissolve without ever having produced precipitation or after a little rain has started to fall from them. Some of them grow naturally and without apparent reason into immense, threatening, towers (cumulonimbus) precipitating copiously. How can the effects of human intervention be assessed with some degree of confidence in such a situation?

Cumulus clouds are usually seeded from aircraft. This is more costly than seeding with ground generators. Since farmers, who are known to be good with figures, year after year employ the services of weather modifiers for working on cumulus clouds, it must be assumed that they do it on the grounds of their experience, irrespective of the fact that a firm statistical proof of success – or failure – of these operations can generally not be produced.

Besides the activities of commercial weather modifiers, there have been a large number of scientific experiments with cumulus clouds and convective systems in the USA, Australia, Israel, the Soviet Union and other countries. These experiments show that seeded clouds produce on average 10–20% more precipitation than unseeded clouds.[48] Even strict statistical tests indicate that these successes are unlikely to be all accidental.

The economic benefit of such operations must vary with differences in local conditions. In Israel, an increase of precipitation from 10 to 15% would already be a significant gain. During experiments with cumulus clouds in the steppe areas of the Ukraine, E. E. Kornienko obtained statistically significant results from seeding of cumulus clouds. However, thorough calculations showed that summer rainfall could be increased by about 1% only with this method, and that the required effort would be too large.[49] He subsequently started to experiment with cumulonimbus clouds and formed the opinion that, on average, this brought increases in precipitation of 35%.[50]. The statistical proof of this has yet to come. Cost–benefit analyses for a target area of 5000 square kilometres, on which winter wheat, barley, maize and sunflowers are grown, have shown that seeding can produce a profit of 0.2 million roubles (2%) in years with good precipitation, 1.5 million roubles (8%) in years with average precipitation and 2 million roubles (24%) in drought years.[51]

While these figures are impressive statistically, there is so far no evidence for operational programmes of precipitation augmentation in the Soviet Union. 'Somewhat surprisingly, considering the extensive and costly droughts that struck the grain belts of the USSR in 1972 and 1975, the overall efforts in rainfall augmentation seem to be quite modest; the investment in rainfall augmentation research is relatively small and apparently there are no operations going on to increase rainfall for the purpose of increasing agricultural production', writes Lewis J. Battan in a report on a visit of American scientists who studied the Soviet weather modification activities by invitation of their Soviet colleagues.[52]

The main seeding advantage of orographic clouds is in the continuous supply of water vapour by the air streaming through them. This water vapour is the raw material for the production of droplets. In cumulus clouds, the supply of water vapour stops after they have reached their maximum development (generally after 1 hour). Thus, a mechanism designed merely to facilitate the 'liquifaction' of water vapour cannot be expected to yield very large increases over the natural precipitation of mature cumulus clouds.

'It is curious', writes Professor E. K. Fedorov, a leading Soviet weather modification expert,

that one figure for precipitation augmentation – about 10 percent – recurs practically unchanged in an enormous number of experiments carried out by different means in widely different regions of the globe. A 10 percent increase of precipitation implies extraction of about 50 to 70 percent of the water contained in the treated clouds occurring over the experimental territory which would not ordinarily produce precipitation by natural means. It must be borne in mind that in natural precipitation the cloud, as the experiments of Soviet researchers have shown, produces approximately 10 to 20 times more water than is contained in it at any given moment; that is, over a period of time it acts like a peculiar kind of water generator, forming the water vapour found in the air into water droplets or crystals, which then fall to the earth. The real object, if we are to obtain a significant augmentation of precipitation is to find some way of stimulating the water-generating properties of a cloud.[53]

A solution to this problem requires that the internal dynamic processes of a cloud be amplified. This had already been recognised by Langmuir; and in 1949, Bergeron pointed out that such indirect effects of cloud seeding had apparently occurred during the first Australian seeding experiments. Later, the renowned American cloud physicist James E. McDonald and other theoreticians dismissed this idea as impracticable. It was only in the course of

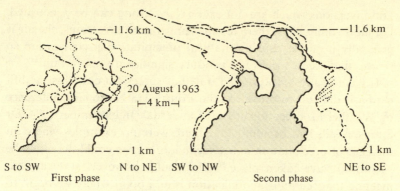

Fig. 15. Explosive growth of a cumulus cloud following dynamic seeding. This sketch is based on a series of photographs taken in short intervals. (After J. Simpson and A. Dennis in Hess (*a*).)

theoretical work on computer modelling of cumulus clouds, that Joanne Simpson and others found the original ideas of Langmuir and Bergeron confirmed. Extending them they developed the concept of 'dynamic seeding'.[54]

Dynamic seeding is a kind of overseeding, whereby a hundred times more seeding material is used than during ordinary operations. Its main objective is not the production of precipitation, but the inducement of cloud growth: this is achieved by the liberation of large amounts of latent heat energy when supercooled droplets change to ice crystals. The dynamic development of the cloud is thus increased and the cloud may grow to heights where it is sufficiently cold for ice crystals to form naturally. Also, because of the stronger updraught inside the cloud, larger amounts of moist air are brought up from lower atmospheric layers and a greater supply of the 'raw material' is maintained.

Successful dynamic seeding induces cumulus clouds to grow 1 or 2 km higher and to produce twice as much precipitation as unseeded clouds. Experience gained during the tests in Florida shows that dynamic seeding is particularly effective on fine days when the so-called inversions exist. On such occasions there is a reversed temperature gradient at a certain level in the atmosphere: it is warmer aloft and colder below. The ascent of air necessary for the growth of the cloud is thereby halted at the inversion layer. Dynamic seeding increases the buoyant forces which push the air up inside the cloud

63

and inversions which are not excessively strong can be penetrated. On moist rainy days, there are no temperature inversions, the air is generally unstable, and the natural precipitation processes are so strong that seeding does not produce significantly more rain.[55]

It has already been pointed out that successful cumulus modification is difficult to prove. At the Fourth US National Conference on Weather Modification at Fort Lauderdale in November 1974, Dr Simpson still had to admit that there were no examples where an increase of precipitation *in a specific target area* could definitely be shown to have taken place.[56] She could, however, state at that meeting that 'cumulus modification is at a point where we begin to know what we are doing...We have finally attained the capability to formulate tractable questions on difficult aspects of convection, ranging from the relations between dropsize spectra, ice concentrations and updraughts through cloud interactions to extended-area effects of seeding.'[57]

One of the main problems occupying the minds of the Florida team at that time was the triggering of so-called 'cloud mergers'. Observation of natural processes had led to the conclusion that the main source of precipitation – at least in southern Florida – is the big cumulus clouds which originate from the merging of two or more smaller cumuli. The crucial need for investigating this phenomenon was recognised, but in 1973 Dr Simpson stated that we were 'no farther ahead in this task than we were in 1945 in treating individual cumuli. And the problem is at least ten times more difficult because it involves the meso-scale (10–100 km) of atmospheric phenomena, virtually unmeasured and unknown.'[58]

Since then considerable progress has been made. The Florida team now claims that it is able to promote mergers by simultaneous heavy seeding of two or more individual cumuli in close proximity and that in this way 'a net increase of rainfall over a fixed target area' can be achieved. It seems that the introduction both of a new kind of silver iodide flares with greater nucleating efficiency and of new small aircraft of higher manoeuvrability has considerably contributed to the success.[59]

Analyses of the Florida results have shown that there is yet another factor which strongly influences the effects of seeding: cloud motion.

64

Dynamic seeding is assumed to promote rapid cloud growth, but it appears that clouds cannot get enough 'raw material' for growing, in the form of moist air from the layers underneath them, when they remain at the same place. When moving, they can feed on fresh sub-cloud air which seems to promote growth. Best seeding successes were achieved on days with light winds at speeds below 6 knots (about 10 km/h).[60]

While the Florida experiments are continuing, a new large-scale investigation (Project HIPLEX), scheduled to last for 5 to 7 years, has been started by the US Bureau of Reclamation to explore the possibility of producing additional rain from cumulus clouds in the more continental setting of the semi-arid high plains.[61]

Rain from 'warm' clouds – and from blue skies?

Cloud seeding with silver iodide and dry ice is only successful when the cloud contains supercooled droplets which can be converted into ice crystals. 'Warm' (not supercooled) clouds, like 'warm' fog, must be treated with rock salt or other hygroscopic substances; this creates considerable technical and ecological problems. Since the seeding with salt does not influence the internal dynamics of a cloud, it can, at best, produce the precipitation of some of the water that is already in the cloud and no more. Seeding with salt has been carried out in southern areas of the USA, as well as in other warm countries. Some of these experiments were moderately successful; others failed or their results are disputed.[62] Experiments in India, where an unexpectedly big success cannot be explained theoretically, have already been mentioned. The spraying of water on 'warm' clouds can trigger a coagulation of small droplets capable of producing precipitation and, on balance, more water is precipitated from the cloud than has been sprayed onto it.[63] The cost–benefit ratio of this technique, however is, poor.

The practical benefits of rain making should not be assessed only by the instantly produced rainfall amounts. In hot countries, droughts may arise as a result of too many condensation nuclei being present in the air. When condensation occurs, many tiny droplets are produced which are too small to fall to the ground. This 'natural overseeding' has an amplifying feedback effect: absence of precipitation → dry surface → much dust → too many condensation nuclei in the atmosphere → lack of precipitation.

The artificial stimulation of even the lightest of rains, which would soon evaporate and make no impression on soil moisture, can wash out many of the condensation nuclei in the atmosphere and for a while pin them down on the moist ground. In certain situations, this is sufficient to break the vicious circle and to produce more suitable conditions for the occurrence of natural rainfall.[64]

But what can be done when there are no clouds? Can they be produced from the clear sky? Several scientists, among them a Frenchman Henri Dessens, answered this question positively. They point out that large fires produce strong convection currents which often cause the formation of clouds. After his first trials in southern France in 1955, Dessens was invited by an organisation of large plantation owners in the then Belgian Congo [now Zaïre] to produce rain. He experimented with controlled burning of scrap wood and later together with his son Jean, with enormous burners, which converted up to 100 tons of fuel oil into smoke within half an hour. These fires often produced clouds, but seldom precipitation.[65] The cost of the efforts far exceeded the benefits.

Dessens's ideas were taken up in the Soviet Union and developed further. Several jet engines were installed side by side and produced a strong vertical current of hot air, said to be capable of breaking through a weak inversion and of causing the formation of large clouds.[66] There appear to be no data for assessing whether precipitation can be produced in this way, or what the cost–benefit ratio is.

At relatively high humidities (above 90%), artificial clouds can also be produced by seeding the air with hygroscopic substances.[67] These clouds are small, dissolve rather quickly, and generally do not grow to the size required for the production of significant rainfall.

Artificial cumulus and cumulonimbus clouds, as well as an increase in the dynamics of the existing clouds can, according to the view of the American geophysicists W. M. Frank and W. M. Gray, be created by seeding of moist air with soot, especially in regions of cyclonic low pressure and fronts.[68] These scientists also believe that seeding air with soot, over moist vegetated ground, can augment evaporation through increased warming and thereby increased cloudiness. No such experiments have yet been carried out. The environmental desirability of such a technique is highly dubious.

On the other hand, clouds (and 'warm' fog) can be dissolved, if desired, by spraying soot on the side which faces the sun; the black soot takes up more heat than the white surface of the cloud.

The purpose of rain making

Augmentation of precipitation is not a purpose in itself. Figures of percentage increases in annual rainfall tell little about the actual benefit of weather modification. More important is the capability of delivering water when and where it is most needed.

There is little danger of ill-timed weather modification worsening a flood, because situations which are prone to produce heavy natural precipitation are not susceptible to cloud seeding, or show only a minimal response. There does not seem to be a chance of a drought being ended, because there are often too few clouds or no clouds at all in the sky which can be influenced by recently known techniques. 'We are only now beginning to study the characteristics of clouds during drought conditions and evaluating the potential effectiveness of cloud seeding during these conditions' reads a report of the American National Science Foundation presented at the World Meteorological Organization Conference in Boulder.[69] The most suitable seeding situations exist on cloudy days with low natural rainfall. How can we take advantage of these particular weather situations to maximise the practical benefits?

Agriculture, as far as it is carried out in nonirrigated areas, is mainly interested in additional precipitation during the plant growing period. Augmentation of rainfall is not always possible when it is most needed, but there are usually some periods when many clouds pass overhead without bringing the desired rainfall. In such cases cloud seeding can be helpful.[70] Moreover, if it is possible to increase precipitation during 'normal' weather situations, the associated increase in soil moisture can be of benefit later during drought periods, when artificial rainfall stimulation is not possible.

A proper cost–benefit analysis for agriculture is very difficult because lack of water is not the only factor limiting crop yield. The effects of artificial precipitation must be evaluated in context with

67

other variable factors (fertilisation for example). Comparing the production cost of a unit volume of water, cloud seeding is considerably cheaper than desalination of sea water by reactors. In practice however, any methods which can produce water *independently of meteorological conditions* may show a better cost–benefit ratio than the cheaper, but not so reliable rain making.

Notable successes have been claimed for applying cloud seeding to combating of forest fires. Cumulus or frontal clouds are seeded as they approach the fire. Since the heat above the fires produces ascending air currents, clouds are lifted in the same way as they would be by an orographic obstacle and conditions suitable for successful seeding are produced.[71] Some 308 forest fires have been put out in Siberia over the period 1971/75;[72] in 214 (77%) the rain fell on to the fire and in 121 of these (43% of the total) the fire was thereby extinguished. In other cases, the artificially produced rain facilitated the work of the fire brigades considerably.[73] This technique is also applied in Australia and Alaska on a routine basis.[74]

Many other potential beneficiaries of weather modification, such as power stations and agricultural irrigation works, use water reservoirs. The crucial question is whether cloud seeding can produce a better utilisation of the reservoirs' capacity. Drinking water reservoirs of large towns usually have a capacity of only a few days' water supply. Under normal conditions, the springs and pumping stations supply more water than is used; so the reservoirs are topped up each morning and partially emptied during the time of peak demand. Only during prolonged droughts does the daily use exceed the inflow. Economy measures, such as a ban on washing cars, must then be imposed. The prospects for additional water supply by cloud seeding are small during such drought conditions.

When the water demand of a town grows faster than the capacity of the water supply installations, however, reservoirs rarely get topped up and water shortages are frequently experienced – as was the case in New York during the end of the 1940s. In such a situation, practically any kind of precipitation augmentation will be beneficial and desirable to bridge the emergency until additional water supply installations come into operation.

Lake Tiberias in Israel is a water reservior of a special kind. Water is pumped out of it continuously, but a certain amount must also flow into the Jordan river. Israeli hydrologists have pointed out to the weather modifiers that an augmentation of precipitation in the catchment area of the lake is only then of practical use when it is *not* in the form of heavy rainstorms. The capacity of the pumping station is limited and sudden strong increases of rainfall in the catchment only creates an excessive overflow of unused water into the Jordan river.[75] What makes cloud seeding particularly attractive in this case is that it is likely to produce a maximum augmentation of rainfall on days when little natural rainfall is expected.

Hydroelectric stations built into dams across large rivers produce power *continuously*. The reservoir capacity is small compared with the amounts of water flowing through them. Such stations cannot fully utilise the water supply when the level is high and part of the water flows unused over the weir. When the river is low the stations work below full capacity. Cloud seeding in the river catchment of the stations is only useful if it can increase the flow rate during times of low water level.

Power stations connected with reservoirs in the mountains serving to supply *peak* power requirements for a few hours during the day are often backed by a volume of water as great as the total annual supply. The reservoir is filled during the rainy season and the water level drops during the drier parts of the year. The construction of large dams is expensive and it is not economical to make the holding capacity of the reservoir large enough to accommodate all the water which becomes available in a year of particularly high rainfall; they are designed to be filled completely during years of low rainfall, and part of the water flows unused over the spillway during years with higher rainfall.

This leads again to the question of whether, and to what extent, rainfall can be increased by cloud seeding during years with *low* natural rainfall, because any additional rainfall produced in other situations can either not be used at all or only to a limited extent.[76] Orographic seeding in the catchment area of a reservoir can produce additional rainfall but amounts will vary from year to year, being generally lower in drought years than in years with abundant rainfall. Thus, for the designer of the power station only the guaranteed minimum amount of rainfall augmentation is of interest. At the start of the season the year's rainfall cannot be anticipated and it may be that precipation is generated which is of no practical benefit to the power station. But an electric grid does not only consist of hydroelectric stations. It can supply additional power from thermal stations during drought. Therefore, in designing the holding capacity of reservoirs the economy of the whole system must be taken into account.

In areas where water *per se* is scarce reservoirs are often found which hold *more* than the year's supply, so that the excesses in a year with abundant precipitation can be saved for when precipitation is deficient. This situation exists in the Colorado Basin, where the holding capacity is about three times the average annual water supply.[77] Here are conditions where artificial increases in precipitation, no matter how much and when, are of benefit. This was one of the reasons for the first large-scale experimental project of precipitation augmentation by orographic cloud seeding having been performed in this river basin.

Cloud seeding as a weapon

In the days of Langmuir, the American military had already become interested in cloud seeding and 'Project Cirrus' was financed entirely from defence resources.[78] According to a summary, published by the US Department of Commerce, on weather modification operations during the financial year 1972, the Naval Weapon Center in China Lake, California, undertook experiments using various techniques of seeding, and also carried out a study of the extended area effects of seeding.[79] Personnel of the US Defense Department carried out a large operation to alleviate droughts in the Philippines.[80] We might ask whether the military took on this task for purely humanitarian reasons, or whether they welcomed the opportunity to try out techniques and thus gain experience which could eventually be used for less humanitarian purposes elsewhere.

One of the first reports available to the general public, concerning the possibilities of environmental warfare, among it the application of weather modification as weapon, appeared in 1968. Its author was the geophysicist Gordon J. F. MacDonald, who, in the same year, had retired from his post as Vice President of the US Institute for Defense Analysis.[81] Until then, MacDonald had been a member of the so-called 'Jason'-group, a body of scientists from various disciplines who occasionally discussed the scientific progress with representatives of the Defense Department. During the 1960s he was chairman of a commission set up by the American Academy of Sciences and charged with the assessment of results and perspectives of weather modification which, at that time, published a mildly optimistic report.[82] It could therefore be assumed that he was well informed about this matter. Regarding the possibilities of cloud seeding for military purposes he writes:

One could, for example, imagine field commanders calling for local enhancement of precipitation to cover or impede various ground operations. An alternative use of cloud seeding might be applied strategically. We are presently uncertain about the effect of seeding on precipitation downwind from the seeded clouds, but continued seeding over a long stretch of dry land clearly could remove sufficient moisture to prevent rain a thousand miles downwind. This extended effect leads to the possibility of covertly removing moisture from the atmosphere so that a nation dependent on water vapour crossing a competitor country could be subjected to years of drought. The operation could be concealed by the statistical irregularity of the atmosphere. A nation possessing superior technology in environmental manipulation could damage an adversary without revealing its intent.[83]

The practical use of cloud seeding was no longer mere theory at the time when the above lines were written. At least since 1967, this technique had been applied in the Vietnam War to increase rainfall over the Ho Chi Minh trail and thus to hinder supplies to the Vietcong. For many years US officials denied any such activity. Even after the publication of *The Pentagon Papers*, which contained a remark on the performance of weather modification in Laos (Operation Popeye).[84] Defense Secretary Melvin Laird stated in April 1972 before the Senate's Foreign Relations Committee that no such operations had ever been carried out in North Vietnam. This was taken as an implied admission that clouds had been seeded above Laos and Cambodia.[85] Regarding North Vietnam, Laird had not told the truth, as transpired from later official statements which showed that, until the end of bombardment of North Vietnam, seeding had taken place over that country too.

The matter was in the end unexpectedly clarified in a macabre way when the Weather Engineering Corporation of Canada and their affiliates in the USA accused the American government of having used a special cartridge developed by their firm for seeding operations in Southeast Asia, thereby infringing the patent law, and demanded payment of $95 million in licence fees.[86]

Not until two years later did the Pentagon change its attitude. During an enquiry before the Foreign Relations Committee of the Senate on 20 March 1974, Lt Col Ed. Soyster, representing the joint Chiefs of Staff, finally presented exact data on the seeding operations in Southeast Asia.[87] The objective of the programme had been to increase rainfall during the transition seasons (at the beginning and at the end of monsoon) in carefully selected target areas in Laos, Cambodia, and initially also in North Vietnam, to make traffic through the jungle more difficult and to trigger landslides along roadways and wash away river crossings. Such events occur naturally during the height of the monsoon rains and the intention was to prolong the time during which the enemy had to face supply difficulties. The seeding apparatus was developed at the Naval Weapon Center at China Lake and commonly known commercial seeding techniques were applied. After initial experiments in Laos, operations started with the consent of President Johnson in 1966 and lasted until July 1972. The seeding aircraft were based in Thailand. The annual costs of the programme ran to about $3.6 million.

The effect of the operations, explained Soyster, could not be quantified. Evaluation was based on assessment by air crew, visual and photographic reconnaissance, and intelligence information. Its impact was measured mainly by the traffic density on the Ho Chi Minh trail which was recorded by automatic sensors. Comparisons of precipitation from seeded and unseeded clouds, as are usually made during scientific experimental programmes, had not been done. As an example for the effects of the seeding programme, Soyster provided detailed data for June 1971. At that time the total rainfall in the affected areas was between 2 and 28 inches (5–70 cm), of which probably 1–7 inches (2–18 cm) could be attributed to the seeding. One of the Senators compared the programme to the labour of an elephant which is giving birth to a mouse. Also present at the enquiry was the Deputy Assistant Secretary for Eastern Asia in the Defense Department, Dennis J. Doolin, who admitted that these results were not particularly impressive.

The programme was considered sensitive and strict secrecy was maintained. Besides the military, only the President, the Defense Secretary, the Secretary of State, the Director of the CIA, and a few of their closest collaborators were kept informed. Doolin himself stated, on being questioned, that he only learned of it afterwards through the press, although he had been responsible for that part of the world in the Defense Department for five years. Neither had the Office for Arms Control and Disarmament been informed, nor the government of Thailand, from where the aircraft operated. The Royal government of Laos, which kept up good relations with the USA, had not been notified of the operations carried out in its country. There was no answer to the question of why there was such a strict secrecy about operations which were allegedly limited to techniques used for tens of years by commercial weather modifiers. After all, the US had used much more atrocious weapons during the Vietnam War.

The question of whether there was any connection between the cloud seeding and the heavy flooding in North Vietnam in 1971 was answered by the Assistant Secretary Doolin as follows: 'The flooding in North Vietnam, as you will recall, generated widespread civilian suffering and that was never the intention nor the result of this programme.'

The statement of the Pentagon speaker caused two kinds of doubts. Scientists, who in general were sceptical about cloud seeding, were of the opinion that the successes claimed by the military – increases in precipitation up to 18 cm – had not been proved and that there was no adequate basis for estimating how much rain would have fallen without the seeding operations. Commentators with less knowledge of meteorology, but a lot of experience with the practices of the Pentagon, put forward doubts of another kind; they were not prepared to believe that the military which continuously inflicted cruel suffering on the enemy's population by napalm, plastic bombs and many other weapons would suddenly have scruples about possible suffering and loss of civilian lives from the application of the weather weapon. The American journalist James W. Canan, who has a thorough knowledge of American weapons and armament plans, writes in this connection:

The Pentagon people point out that cloud seeding had been the object of civilian research and development for many years, and that the military had simply found it compatible with the cause of the war. What they did not emphasise was that the 'technology' has now proceeded to the point that not just rain showers but torrents can be triggered, that entire continents can be targeted for catastrophic cloudbursts.[88]

In a generally well-documented book, Canan does not provide source material or proof for this statement and many meteorological experts are likely to doubt that weather modification on such a large scale is possible at all. On the other hand, if it were possible to trigger devasting weather catastrophes, this would be a terrible weapon even if only one out of ten attempts succeeded. There would only be the question of how to find out whether the effect was really caused by military weather making and not just by chance.

As far as can be judged from this book, Canan is a serious journalist, who apparently had good relations with the Pentagon and reported for many years on the activities of the military. He describes numerous weapons which already exist, which are in the development stage, or which are on the drawing board – weapons even more cruel than those which trigger catastrophic weather on a continental scale. Weather modification is treated in his book only as a side issue, without special emphasis or sensationalism. It is of course possible that one of his military contacts intended to mis-inform the public in this way, or that the efficacy and reliability of weather as a weapon is overestimated by the Pentagon. But it is also conceivable that the US military – and perhaps those in other countries – have carried out secret research of which civilian meteorologists have no knowledge, and that they were interested in practical trials of the results of such research.

If the Secretary of Defense, Laird, did not tell the truth when he said that cloud seeding had never been carried out over North Vietnam, why should it be taken for granted that Doolin, the Deputy Assistant Secretary in the Department of Defense, spoke the whole truth when he denied a connection between cloud seeding and flooding? Vietnam was a proving ground for many weapons and war tactics without regard to the human suffering thereby inflicted. The possibility of testing meteorological warfare other than making

74

unpassable paths and roads in the jungle by precipitation augmentation cannot be excluded.

Hail prevention – more practice than theory

According to American estimates, world agriculture loses about $1000 million annually through hail damage.[89] Additionally, millions of dollars worth of damage are caused to nonagricultural assets. Personal injuries, such as occurred in July 1977 in a French holiday camp, are not often reported from developed countries, but in a country like India it can happen that a single heavy storm causes hundreds of deaths by hail.[90] It is therefore not surprising, that there have been longstanding attempts to reduce or prevent hail. Hail shooting with mortars has been practised for a long time to protect vineyards and orchards. In 1896, the Austrian Albert Stiger tried out a 'hail cannon' in an orchard region of Styria. This was a 3-cm mortar which had the funnel of a steam locomotive mounted above it. The gadget made a deafening noise on firing and produced a ring of smoke which rose into the air with a singing noise. After allegedly successful initial results, the use of this cannon spread rapidly through central Europe. Its practical application led to repeated accidents and in 1902 the Austro–Hungarian Government organised an international conference to evaluate this technique. This was followed by scientifically supervised experiments in Austria and Italy, as a result of which the cannon was pronounced entirely useless.[91]

Shortly after the Second World War, the French General F. L. Ruby developed an inexpensive small rocket of limited altitude range, carrying explosives which a farmer could fire at a threatening cloud. H. Byers recalls that this led to a phenomenal business. The psychological effect of thus being able to shoot at the enemy 'must have been very satisfying'.[92] The practical effects, however, were not so impressive. As a matter of fact, there was none at all.

Between 1948 and 1952, Swiss meteorologists carried out an experiment with explosive rockets in the Tessin (Large Scale Experiment I) and found no difference between protected and unprotected areas. In Kenya, where an area climatically favourable for planting of tea, suffers continually from heavy hail, more than 10000 shells were fired against 150 heavy thunderclouds during a period of four years. The alleged success could not, however, be scientifically proved.[93]

As cloud seeding techniques developed, it was natural to apply them also in hail prevention experiments. American commercial weather modifiers soon developed a fitting 'theory': through 'over-seeding', the formation of large and dangerous hailstones was supposed to be replaced by the production of much larger numbers of smaller hail pellets which cause less damage, or even melt before they reach the ground. Cloud seeding against hail was soon taken up in many countries and there was no shortage of some rather impressive success reports. But even now, after a quarter of a century, it is still debated whether, and to what extent, the successes claimed to have been achieved by this technique are real; and there are good reasons to assume that the 'theory' on which these techniques are based is incapable of describing the actual processes of hail formation in a thundercloud.

Reports of success generally come from operational programmes which do not include simultaneous control experiments for the purpose of statistical comparison. Successes are usually derived by comparison either with hail losses in earlier years or with hail losses in unprotected areas. The conclusiveness of this is doubted by many scientists, especially for hail which fluctuates naturally over a large range in location, time and intensity.

Nevertheless, however justified such doubts may be in individual cases, and in spite of the fact that some commercial hail protection programmes have completely failed – as in the San Louis Valley, which was mentioned at the beginning of this book – the number of successes reported is so large that they cannot simply be swept under the carpet. Since success of hail shooting with explosive shells could not be proved, the tea growers of Kenya contracted an American weather modification company which uses the cloud seeding technique. Since then, hail damage has allegedly been reduced by 70–80%.[94] The meteorologist Stanley Changnon, who had declared at the Conference of Tashkent in 1973 that all North American programmes so far carried out 'have not as yet provided solid evidence that hail can be suppressed, at least to the satisfaction of a majority of scientists', has meanwhile published 'best estimates' of results of more recent operations which are rather impressive.[95]

In four of the larger US programmes – among them the state-wide project of South Dakota, in which, at times, up to two-thirds of the area of this state participated – and in a South African programme, reduction of damage between 20–50% has been achieved.

By far the biggest success claims come from the Soviet Union. There, hail prevention programmes extend to an area of about 5 million hectares and damage reductions of 50–90% are reported.[96] Silver iodide and lead iodide are carried by big rockets into the suspected hail formation zones of thunderclouds. Particular techniques vary slightly in detail from place to place, but substantial success is reported from everywhere. Strict scientific attempts to assess the efficacy of the applied techniques have never been carried out in the USSR. Yet the mere size of the protected areas seems to preclude the possibility of purely accidental successes. 'As long as the operations extended only to some ten thousands of hectares', writes Professor Federov,

everyone understood that no serious conclusions could be drawn from this, since the probability of hail falling on precisely that small area was very slight. Only after we had protected an area of up to a million hectares, and worked on it for two or three seasons, could we estimate the effect of modification with any degree of confidence, and persuade ourselves that immense sums of money would not be spent in vain.[97]

It is obvious that such massive success reports impressed scientists and weather modifiers in other countries and made it difficult for them to admit negative results 'when others are claiming such drastic positive ones'.[98]

Scientific attempts to prevent hail have generally been much less successful than operational programmes. Experiments are randomised: whenever the danger of hail arises, whether seedings should be carried out or not is determined by throwing dice or by any similar random procedure. The results on seeding days are then compared with those on control days.

Of two relevant experiments in the Swiss Tessin, where ground generators were used, the first (Large-Scale Experiment II) had to be terminated prematurely, since it became obvious that its layout did not allow a proper statistical assessment of the results. However, the second (Large-Scale Experiment III, 1957–63) clearly showed that hail damage had *not* been reduced by the seeding and that it may even have been increased on some occasions. Rainfall amounts, however, had been augmented. On days with pronounced southerly winds, this effect could even be shown on the leeward side of the Alpine mountain range, at Bern and Zürich.[99] Swiss meteorologists have come to the conclusion that seeding by ground generators is not a suitable means of reducing hail in the Tessin area.

Also in Southern France and in the Argentinian wine-growing area of Mendoza, the results from experiments with ground generators were disappointing.[100] Subsequent analysis of the Argentinian results showed that this kind of hail prevention technique was effective after the passage of cold fronts, while seeding in other meteorological situations not only had little effect but possibly even increased the hail damage.[101]

Following the persistent reports of success by weather modifiers in America and in the Soviet Union, a large scale National Hail Research Experiment (NHRE) was put into operation during 1972 in the United States. Cloud seeding from aircraft was the basic technique used in the operations. The experiment was terminated prematurely after four years when it became obvious that continuation would bring no significant statistical results.[102] Sometimes an increase and sometimes a reduction of hail had been observed after seeding. Besides, cloud physics studies carried out within the framework of this programme had convincingly shown that the planning of the experiment was based on theories of hail formation which were incorrect. The vice-chairman of the NHRE project, R. W. Sanborn, explained in a letter to the present author that a continuation of the seeding experiments should be considered only after 'intensive research of the dynamics and microphysics of severe convective storms'. Such research addresses a basic problem, which has so far bedevilled all attempts to develop a scientifically based hail combat technique: lack of knowledge of the natural process of hail formation.

The technical difficulties associated with hail research are considerable. Flying an aircraft into a large thundercloud to investigate the conditions at the spot had, for a long time, been considered too dangerous. Only recently have such flights been carried out in the

United States with specially armoured aircraft, and valuable results have been obtained. Previously there was only less direct evidence from radar observations, examinations of hailstones and laboratory experiments. At present, little is known about the actual number of natural icing nuclei and hail embryos in different types of thunderclouds, and about their relative numbers compared to the amount of liquid supercooled water in the clouds. Almost nothing is known about some apparent relations between electrical processes in the cloud and the formation of precipitation. Also, the processes which govern the growth of the hail from the embryo stage to a size that makes the stones a real danger are not completely understood. It has been found that different hailstones have a different structure: in some, the core is made up of clear ice, in others, of opaque soft ice. The latter apparently had a 'dry' growth, meaning that rime was deposited on the hail embryo. Clear hailstone embryos grow 'wet' – liquid water builds up on them and subsequently freezes. Some hailstones are uniform, others consist of different zones and layers which can be seen with the naked eye.[103] Occasionally hailstones with different structures fall from the same thundercloud. This suggests that there are different regions in a cloud where hail can form under different conditions. Radar studies and meterological observations have shown that large-scale structure and temperature conditions differ between clouds and within one cloud.

In view of such a variety of natural processes, it is difficult to see that there could be one generally valid recipe, such as overseeding, which would produce a reduction of hail in each case. In fact, American scientists have developed a number of plausible concepts as to how seeding can trigger an increase rather than a reduction of hail in certain meteorological circumstances.[104] So far, all these are hypothetical ideas which have to be investigated. They do, however, provide a possible lead to understanding why so many scientific experimental series give contradicting results. Seeding 'blindly' without sufficient knowledge of cloud physical processes seems to have a different effect under different conditions, leading sometimes to an increase and sometimes to a decrease of hail. Separation of the different circumstances with hindsight is often not possible since the necessary meterological data were not collected during the experiments.

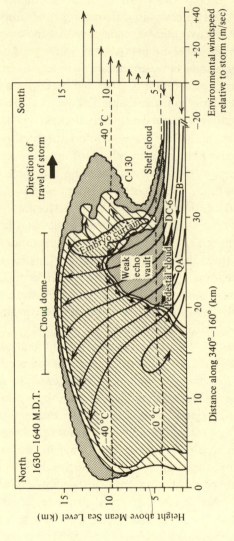

Fig. 16. Vertical section in a plane oriented from NW (340°) to SE (160°), showing features of the visual cloud boundaries of the Fleming, Colorado, supercell storm at 16:30 to 16:40 MDT, 21 July 1972, superimposed on the pattern of radar echo. The section is oriented in the direction of travel of the storm. Two levels of radar reflectivity are represented by the different densities of hatched shading. Areas of cloud devoid of detectable echo are shown stippled. Short thin arrows skirting the boundary of the vault represent the hailstone trajectory. The thin lines are streamlines of airflow relative to the storm drawn to be consistent with the other observations. C-130, DC-6, QA and B signify positions of aeroplanes. To the right of the diagram is a profile of the wind component along the storm's direction of travel, derived from a Sterling, Colorado, sounding 50 km south of the storm. (After K. A. Browning and G. B. Foote, *Quart. J. Roy. Meteor. Soc., Lond.*, **102** (1976), 499.)

Seen in this light, hail modifiers are today in a situation probably similar to the one the rain makers were in before they became aware of the limits to the temperature range within which cloud seeding provides an increase in precipitation. There is yet another handicap in hail modification: the natural processes which are being interfered with are probably far more complicated than those of other precipitation formation and, accordingly it might prove much more difficult to come up with reliable 'seeding recipes'.

An example of the kind of problems that arise is assessing the chances for successful seeding in the so-called 'super-cell storm'. This is a mature large-scale thunderstorm which can move across country for several hours without a significant change in the structure of its cloud system; the most severe hail damage probably comes from this type of thunderstorm. It is particularly important to try to find a way of modification that can avert this danger. Fig. 16 shows the picture of such a cloud as it appears on the radar screen. At the front of the cloud is a zone which hardly reflects the radar beam (weak echo vault), and which consists of minute droplets and crystals hardly visible by radar. Here, the updraught is so strong that condensation and freezing nuclei move very rapidly and there is insufficient time for larger drops and crystals to form. Ahead of this echo-free zone is the 'embryo curtain' which can be picked out clearly on the radar screen. It is assumed that the curtain is made up of ice particles which originate from the relatively quiet region of the echo-free zone and which have grown into soft hail. As soon as a certain size is reached, the soft hail begins to fall. Part of it is entrained into the strong updraught and moves along the path shown by the small arrows. Since there is a large amount of undercooled water in this region of the cloud, the particles can grow rapidly and form large hail stones.

Seeding of the region of strong updraught, which corresponds to the original 'theoretical' ideas of American commercial weather modifiers, was tried in the NHR Experiment. According to the view of the American scientists K. A. Browning and G. B. Foote, this offers little hope for success, at least in the case of the clouds in Colorado where the NHRE was carried out. Here the supply of supercooled water is so large that the amounts of silver iodide which one can realistically introduce are simply insufficient to act on all the water in the cloud. The addition of artificial icing nuclei can therefore cause an increased crystal influx into the embryo curtain, thereby increasing the total number of falling hailstones without their size being reduced.[105]

According to the ideas of Soviet hail prevention experts, the seeding material must therefore be brought in large amounts directly into the active zone of hail formation, and also into the embryo curtain, by means of rockets or grenades. Only by this means can all liquid water be frozen in a time short enough to ensure that ice crystals do not have time to grow into soft hail and serve as embryos for larger hailstones. The efficacy of this method is now being tested in a large-scale scientific experiment in Switzerland, planned to run for several years.

Swiss test Soviet hail rockets

Like the American commercial weather modifiers, Soviet hail combat experts started their operations without adequate theoretical basis. At the 1973 WMO Conference, the Soviet cloud physicist J. Sedunov reported of 'long years of varying experiments, mostly based on physical intuition, which helped *in an empirical way* to find optimum methods of fighting hail'. Results

'have proved to be satisfactory enough for practical application to have followed on a wide scale. Nevertheless it is quite probable that the initial assumptions will have to be considerably modified in the light of further knowledge...In the absence of good physical-mathematical models, we have to use numerical schemes which barely resemble the real ones. We have no clear notion, particularly from the quantitative point of view, of many of the processes occurring in the cloud and practically no direct measurements of physical parameters are made inside the cloud.'[106]

At the core of the Soviet theory – or working hypothesis – is the idea of bringing large amounts of silver iodide, as fast as possible, into those parts of the cloud which show a particularly strong radar echo. Details of the reasoning behind how this introduction produces the desired effect are disputed. Even the doyen of Soviet hail prevention, the Georgian scientist G. K. Sulakvelidze, concedes that his model 'does not lay claim to great rigour in describing isolated details of the mechanism of hail formation' and admits that failures in some of the prevention operations resulted perhaps not just from technical and human shortcomings during the performance but also from misconceptions.[107] Scientists in the Hydrometeorological Institute at Leningrad explained openly during the conference at Tashkent that they disagreed with the assumption that hail can be prevented by overseeding. They were of the opinion that another effect of the seeding, namely the changed interior dynamics of the cloud, produces a reduction of hail in many cases.[108]

All Soviet experts agree, however, on one point: the methods applied are effective in practice, no matter if and how the success may be theoretically explained. This is the very point about which most of the Western experts are sceptical. They are of the opinion that the Soviet success reports are based on data which have been collected with unsufficient statistical rigour.[109]

The numbers of personnel employed in hail protection in the Soviet Union are considerable. According to Sulakvelidze, one hail protection unit which looks after about 100000–120000 hectares, consists of 50–60 persons in the Observatory (meteorologists, radar technicians, communicators and auxiliary personnel) and of either 5–7 cannons or 10–12 rocket installations, complete with operating personnel.[110] The American delegation that visited the Soviet Union in 1976 was informed in the Republic of Moldavia that protection of 730000 hectares requires 700 personnel, 400 of them permanent and 300 part time. Altogether, according to the estimates of this delegation, there are 3500–5000 personnel employed in hail protection in the Soviet Union.[111]

Dr A. Kartsivadze of the Georgian Institute for Geophysics told the American delegation that the cost of these hail protection operations was 5–7 roubles per hectare, the benefit being approximately twice as large. The American delegates were told during their visit to the Trans-Caucasian Hydrometeorological Intsitute that the cost of protection operations in an area of 350000 hectares is about one million roubles, the benefit being about 3–6 million roubles. Dr M. V. Buikov of the Ukrainian Hydrometeorological Institute spoke of a cost–benefit ratio of between 1:2 and 1:8. The leaders of other hail prevention schemes mentioned ratios of 1:8 to 1:10.[112]

Soviet techniques of hail prevention have also been applied in Bulgaria, Hungary, and Yugoslavia. According to the report of a Bulgarian scientist at the Tashkent Conference, the following problems arose: insufficient numbers of trained experts (which is not surprising in view of the high numbers of personnel required by the Soviet method), insufficient numbers of special radar sets for observation of cloud, larger air traffic density frequently precluding operations, and, above all, the need to adapt the Soviet model to the local circumstances. Also, the fact that hail rockets were imported from the Soviet Union had, according to the Bulgarian experts, increased the costs over what they would have been if the rockets were produced in their own country.[113]

Whether hail protection operations influence precipitation amounts has not been explicitly investigated in the Soviet Union. The American visitors were told that most of the areas protected against hail do not suffer from drought, and that therefore little attention was given to this problem.[114] The American programmes examined by Changnon contained two cases of rainfall increases by 23% and 7%, respectively, one case with no change, and one with insufficient data. The South African hail prevention programme caused a decrease in rainfall by 4%. The NHRE which proved ineffective as a hail prevention programme caused an increase in rainfall of 25%. The Swiss Large-Scale Experiment III, although ineffective against hail, also produced an increase in rainfall.[115]

It is doubtful that a generally valid rule of thumb can be found which tells if and under which circumstances hail prevention operations will reduce the rainfall. This problem will have to be studied separately for each area selected according to the local geographical and meteorological conditions; the same applies to the implications for the cost–benefit ratio. Conditions will often differ from those in hail protected areas of the Soviet Union where adequate soil moisture seems to be available. American calculations show that on the high plains in the United States only 5% less rainfall would far outweigh the benefit which would accrue from a reduction of hail by 20%.[116]

The American scientists who carried out a thundercloud study during the NHRE are of the opinion that the Soviet model concepts of hail formation processes are incorrect, and that the Soviet techniques would not be successful if applied in Colorado.[117] Soviet weather modification experts counter that seeding from aircraft is far less effective than the use of rockets and shells, and that a combined use of the latter would also work in the USA.[118] Swiss meteorologists,

84

who have dealt with three large-scale experiments and numerous theoretical and laboratory studies over the last 30 years, think that it would be promising and worthwhile to test the Soviet techniques by experiments which conform to the requirements of scientific objectivity.

This is not difficult since the Soviet technique is based on an exactly defined quantitative seeding criterion and does not depend on feelings or intuition for deciding when and where seeding should take place. The Swiss have, in collaboration with French and Italian colleagues, started Large-Scale Experiment IV, designed to clarify whether a hail prevention technique strongly resembling that of the Soviets would work in the Alpine region.[119]

The experimental area chosen is in the Napf region near Lucerne, where air traffic is sufficiently small not to disturb the experiment. An average of 45 thunderstorm days a year occurs in this area, 24 of them with hail. Over the planned five-year experimental period about 120 hail days can be expected. During half of these days there will be seedings, the other half serving as control days for statistical comparison. The selection of seeding days is determined by random method (throwing of dice). 'We have put great emphasis on an early publication of the statistical methods to avoid a bias which applies to many weather modification experiments' writes Dr B. Federer of the Laboratory for Atmospheric Physics of the Eidgenössische Technische Hochschule (LAPETH) in Zürich, the director of the Large-Scale Experiment IV, in a letter to the present author. 'This bias arises when statistical methods are not agreed upon before the experiment starts and when all kinds of statistical techniques are applied until one is found which produces significant results.'

To prepare these experiments, meteorological measurements were made from 1974 to 1976, in the selected target area. These showed that the Soviet seeding criteria predict hail danger fairly accurately. Computation of hail probability is based on a mathematical model derived from measurements of cloud height, strength of the radar echoes from various parts of the cloud, and temperature at various altitudes. After some initial training, Swiss meteorologists could perform these calculations within a few seconds. During the experimental measurements of the period 1974–76, it turned out that all hail which fell in the target area could be predicted by the Soviet

85

seeding criteria. Occasionally a false alarm arose from this method, when very heavy rain but no hail would fall, or when the hail had melted before it reached the ground. The Swiss then decided to make the criterion a little less stringent. They accept the risk of occasional slight hail damage by not taking any preventive action on marginal days when conditions indicate either heavy rain or some hail. By this saving, the cost–benefit ratio – which is after all the most important issue – can be improved considerably. According to the view of the scientists in LAPETH, any clouds, which are likely to produce heavy hail can be identified without difficulty.[120]

When American experts visited the Soviet Union, cloud physicist Dr J. Sedunov explained that 'he did not regard Large-Scale Experiment IV as a reliable test of the Soviet procedures because, in his view, the methods to be used in Switzerland for deciding when and where to seed will be different from those in the USSR.' He said that in part this was because 'the Swiss will be using different types of radar from the types used in the Soviet Union'.[121] Dr Federer rejects this reservation. He points out that he and some of his collaborators are in possession of certificates signed by Soviet hail protection experts according to which they are able 'to carry out the Soviet hail defence scheme independently'. Moreover, in Switzerland 3-cm radar is used to decide when and where to seed, just as in the Soviet Union.[122]

The official opening of the seeding experiments took place on 9 August 1976, in the presence of two Soviet experts, but during that year there were no more days on which the seeding criterion was fulfilled. The experiment is scheduled to continue until 1980. Initial estimates will only be published when statistically significant material is accumulated.

Chaff against lightning

In the USA about 600 people are killed annually by lightning, and another 1500 injured. Although the country frequently experiences severe hurricanes, more casaulties are currently caused by lightning strike than by hurricanes and tornadoes together. The damage to property by lightning is estimated at several hundred million dollars a year.[123] There are no worldwide statistics on lightning

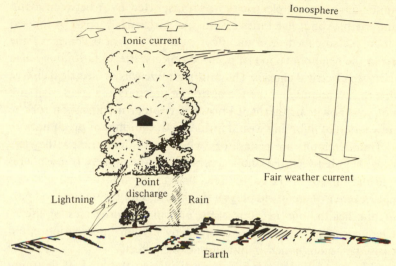

Fig. 17. Schematic diagram of the atmospheric electric circuit. (After G. Dawson in Hess (a).)

damage, but it is known that the number and intensity of thunderstorms in the tropics is much greater than that in moderate latitudes. Thus it can be assumed that in tropical countries damage from lightning occurs more frequently than in the USA.

The combating of thunderstorms is the youngest branch of weather modification. Only a few experiments have been carried out, so far, generally with encouraging results. The theoretical basis for this activity is still rather weak, chiefly because the natural thunderstorm processes are not well understood.

A thunderstorm is a kind of electric generator. There are several dozen theories as to how the generator works to charge the cloud,[124] but it is beyond the scope of this book to discuss them. The charge is probably a result of several mechanisms, each of them of varying relative importance in different clouds. In most thunderstorms the upper part of the cloud acquires strong positive charge, the lower one negative. Lightning flashes are electrical discharges, either between the two 'poles' of the cloud, or between unlike poles of two clouds, or between the cloud base and the surface of the earth. Only under certain not fully understood conditions flashovers occur between parts of a cloud and the ground.[125] Lightning is spectacular but only minor quantities of electricity are transported by it between cloud and earth. About five times as much electricity is carried by ionised air in the so-called 'corona effect', i.e. an electrical charge of the air in the neighbourhood of pointed objects. Precipitation may also carry an electrical charge. The mutual influences of electrical charge in a thundercloud and the formation of precipitation (rain, hail) are not yet known; neither is it known if weather modification for the prevention of lightning would influence the amount of precipitation.

Thunderclouds are generators which continually renew the electrical field between the outer layers of the atmosphere (ionosphere) and the surface of the earth. This field is an important part of our natural environment and plays a significant but little understood part in our health, our nervous state and in the processes of life.[126] Changes in the natural electric field by human intervention could easily be accompanied by unwanted side effects.

This apprehension applies in particular to the kind of lightning prevention which tries to eradicate the problem by prevention of thundercloud formation through overseeding. Relevant experiments (Project Skyfire) have been carried out by the Forestry Section of the US Department for Agriculture several times during the 1960s. The results of a relatively small sample show that lightning occurs only half as often in seeded thunderclouds as it does in unseeded clouds; the number of lightning strokes between cloud and ground is even more reduced. Of particular importance is a result which shows that the duration of lightning flashes from seeded clouds to the ground is markedly shorter than that from unseeded clouds. This observation is very important since it is the relatively long-lasting lightning which causes fires.[127] After a thorough theoretical assessment of the results obtained so far, the seeding experiments of Project Skyfire are scheduled to be taken up again on a considerably larger scale. At the WMO Conference at Tashkent some reports were presented on similar seeding experiments in the Soviet Union, also with encouraging results.[128]

Another way of preventing lightning relies on increasing the conductivity of the air; this allows a continuous transport of electric charges, and tensions are thereby kept from growing too high. To do this, aluminium covered nylon strips, a few centimetres long, called 'chaff', are put into the air. Originally this metallic chaff was used for military purposes to disturb enemy radar during air raids. Lightning prevention experiments with chaff began in the USA in 1965. The metal-covered strips generate corona discharge in the electric field. The two ends of each strip give off opposite electric charges to the air and make it a better conductor. Corona action becomes effective at about one-tenth of the voltage that causes lightning strike. The chaff can either be injected below the thundercloud where it amplifies electricity transport between cloud and the ground, or into the thundercloud where it breaks down the insulation between the two poles of the generator. The first method has the advantage that only the form of electricity transport is changed while the amount of transported charge remains unaffected. This minimises the danger of environmental damage by disturbance of the electric field between the ionosphere and the ground.[129] At present, this method is being tried out in a well-known thunderstorm corridor in Florida and it is reported that initial results are encouraging. Similar experiments are also being carried out in the Soviet Union. There, chaff made from a carbon substance is used which is biodegradable and supposedly less damaging to the environment than the metal strips.[130]

Another method of lightning modification was accidentally discovered in 1968 during the launching of Apollo 12. The space rocket was struck twice by lightning when it moved through a cloud not considered to be a thundercloud and which would not have produced lightning naturally. Subsequent statistical examination showed that lightning strikes aircraft more often than would be expected accidentally. Obviously, metallic bodies in the air can trigger lightning in electric fields at a field strength which would not produce lightning naturally. In this case also, the corona effect appears to have played a significant role. Theoretical considerations as well as practical experiments have shown that it is not so much a question of the flying object's size, but of its shape. A slim pointed rocket only 120 cm long (type Mighty Mouse) has the same triggering effect as the large Apollo spacecraft. Advantage can be taken of this by launching a number of cheap small rockets before the launching of a large rocket or a spacecraft. The specific purpose of the small projectiles is the reduction of the electric field strength so that the spacecraft may be launched without the risk of being hit by lightning.[131]

The first hurricane modification experiment was carried out by Langmuir in 1947. After the seeding, the hurricane suddenly changed the direction of its track. Experts are today unanimously of the opinion that this was by pure accident and had nothing to do with the seeding. But the American authorities have learned a lesson from this incident which has led to a strict regulation: there must be no experimentation on any hurricane that could conceivably reach the coast of the United States. At the present state of knowledge nobody can be sure whether seeding reduces the intensity of the storm – the very objective of the experiment. It cannot be ruled out that seeding might even increase the wind force and, if a storm which had been experimented on were to cause damage, weather modifiers and authorities would be held responsible.

Once this rule was established, it meant that only very few guinea pigs were available for hurricane modifications. Any experimental hurricane must be far out over the ocean and its movement must not be directed towards the continent; on the other hand it must be within endurance range of aircraft specially equipped for such an operation. Several years may pass before a single opportunity for experimentation arises. Between 1961 and 1974 modification was carried out on only four large hurricanes. The results were encouraging, and in one case very much so, but such a small sample does not allow far-reaching conclusions.

Hurricanes and typhoons – names for the same kind of tropical storms in the Caribbean and Southeast Asia, respectively – are long-lived wind vortexes of great strength which form under little-known conditions over the tropical oceans. There is an 'eye' in the centre of the storm measuring about 20 km in diameter. Inside this eye prevail light variable winds and very low air pressure. The low pressure is a result of a strong warming of air in the interior of the cyclone. There, massive condensation of water vapour takes place, whereby large amounts of latent heat are liberated. The water vapour evaporates from the warm ocean surface and is concentrated inside the storm by air flowing spirally towards the centre. More condensation of water vapour generates stronger winds which, in turn, concentrate more water vapour. This regenerating mechanism can maintain a storm of high intensity for several weeks as long as it feeds on sufficiently warm ocean surfaces. When a storm moves over a continent or colder water, it rapidly declines and dies out.[132] The highest speeds of the hurricane winds occur near the storm centres, in the 'wall', at the perimeter of the calm eye; velocities of 200 km/h are not rare and occasionally even 400 km/h have been observed. The circle of hurricane force winds (more than 100 km/h) has a diameter of 100–150 km. Besides direct wind damage, hurricanes also produce storm floods at the coasts. Wave heights triggered by a hurricane over the open ocean can reach 15 m. Near the coast, the waves are somewhat less high, but it can nevertheless come to a catastrophic rise of the coastal sea level of up to 8 m, and even more in bays and estuaries. The intensity and extent of these storm floods depends much on the shape of the sea bed and of the coast. Where water is deep near to the coast, the waves are higher but the rise of the storm flood is less.

When a hurricane crosses the coast and moves inland, very heavy rainfall occurs, especially in hilly areas. Many reservoirs and springs in tropical and subtropical areas are charged by the downpours of tropical storms. They constitute an important and necessary source of water supply. On the other hand, the same cloudbursts cause devastating floods. Hurricane 'Agnes', which in 1972 crossed the coast of the United States in a declining development stage, caused no significant storm damage but triggered the highest river flood levels in the history of the United States, with a record damage of over $3000 million.[133]

The number of human casualties caused by hurricanes has been strongly reduced since weather satellite observations enabled the perfection of early-warning systems and the evacuation from threatened areas. At the same time, the extent of material damage has been rising in recent decades – simply because the living standard of the population has risen and there is now a greater wealth of objects that can be destroyed.[134] Seen from the socio-economic angle, hurricane modification faces a difficult dilemma. On the one hand, it is meant to reduce damage from excessive winds, and flooding, and on the other, it must not impair the vital precipitation provided by the storms that run inland, as droughts can hit the economy of a country more than all the wind and flood damage taken together.[135] The deflection of the storm onto a track which does not reach the coast would – if it were possible – not be desirable. It would be equally undesirable if the formation of hurricanes were prevented altogether. Rather, the storm should be modified in such a way as to lessen the impact of coastal flooding and to redistribute the precipitation over a larger area, thereby reducing the danger of river flooding.

It will probably take a long time to develop techniques which produce the desired effects. In the experiments performed so far, only a redistribution of energy by massive seeding in the outer regions of the vortex has been attempted. This was done primarily to enlarge the eye and thus reduce the wind velocity in the eye wall, where speeds are highest. An important precondition for these experiments is the development of special rockets by the US Naval Weapon Center at China Lake. These provide the possibility of seeding with adequate amounts of silver iodide within relatively short periods.

The technique was successfully applied to hurricane 'Debbie' in August 1969. Repeated short applications of seeding reduced the gust speeds in this storm in 5 hours from 180 to 125 km/h. Two days later, when the wind speeds had again increased to 185 km/h seeding was followed by a reduction to 155 km/h within 6 hours. The probability that the decreases on both occasions were accidental is very small.[136]

If such a decrease of wind force could be achieved in a hurricane which crosses the coast the storm damage could be appreciably reduced. It must be kept in mind, however, that only the top velocities in the eye wall would be reduced by the seeding, not the total energy of the storm. The latter might even be slightly increased. Considering the aspect of wind damage alone, a relative increase of wind speed in the outer regions of the vortex, where the winds do not reach hurricane force, is tolerable. The implications for storm flood damage at the coast are not understood, however, and an assessment is difficult because of local circumstances which differ from case to case. Whether there is a danger of the reduction in storm damage being outweighed by increases in storm flood damage is currently being investigated with the help of computer models.[137]

Since the successful experiment with hurricane 'Debbie', Project Stormfury had only one more guinea pig which was already a bit aged and rickety: hurricane 'Ginger' in September 1971. The seeding of this dying storm brought – as was more or less expected – no significant results.

92

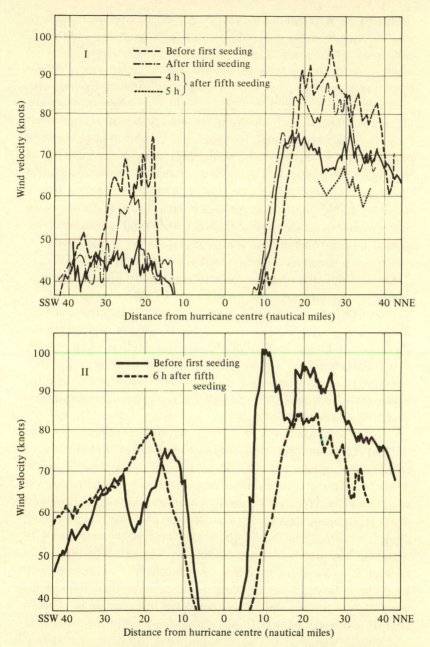

Fig. 18. Change of wind velocity at 3600 m altitude in hurricane 'Debbie' after seeding on 18 August (I) and 20 August (II) 1969. 1 knot = 1 nautical mile per hour (1 kt = 1 nm/h). (After R. C. Gentry in Hess (a).)

Having to wait for many years without an opportunity for a sensible experiment is a frustrating state of affairs. The authorities concerned with Project Stormfury, the Department of Commerce responsible for the US Weather Bureau and the Defense Department have therefore considered shifting their experimental grounds into the Pacific Ocean, possibly to the island of Guam where more guinea pigs are available. Japan, which derives much of its water supply from typhoon-associated rainfall, has expressed its objection to typhoon modifications in the Western Pacific. Moreover, if the influence of seeding is to be judged with confidence, work has to be done on storm vortexes which are beyond the initial development and within a relatively stable phase. The typhoons around Guam do not often have this property; they are frequently in the initial formation stage, sudden natural fluctuations are frequent, and it would be rather difficult to separate accidental effects from those of seeding.[138]

An effect similar to that produced by seeding with silver iodide can, according to some scientists, be achieved by the sowing of soot in a cloud-free region within the area of origin of a tropical storm.[139] This would produce clouds away from the storm and divert the liberation of latent heat from the area of highest wind speeds. Theoretical studies on a computer model of a hurricane predict that this would lead to an enlargement of the eye and, at the same time, reduce the wind speeds in the eye wall. Relevant practical experiments have not yet been carried out.

Another technique capable of reducing the overall activity of the tropical revolving storm (not only the wind force), and perhaps of causing its disintegration, is based on reducing evaporation from the ocean by covering the sea surface with a thin layer of oil or chemical substance. This appears to be promising only when storms are in their state of formation while the sea surface is not excessively agitated and there is a chance that the oil film is not too rapidly mixed up with the water. The area to be treated would be of an extent of 100 square kilometres and a film only a few thousandths of a millimetre thick would be sufficient. Such spraying would require about 1000 hecto-litres of oil – an amount which could be transported in modern aircraft without difficulties. Since the central area of the depression is gradually displaced, the cover would have to be renewed at four to five-hourly intervals. Initial experiments with some sprays (linolacid, polyvinyl alcohol and others) made off the coast of Florida in moderately agitated seas produced a remarkable decrease in wave motion. This would enhance the persistence of the covering layer. An experiment under real storm conditions is planned.[140] The environmental aspects of this method would have to be looked into very carefully because the oceans of the world are already being polluted by too many toxic substances.

The American geophysicist Gordon J. F. MacDonald thinks that a clever combination of the oil-film method with targeted seeding could change the direction of the hurricane track and thus be used as a terrible weapon. 'At present' he wrote in 1968, 'we are a long way from having the basic data and understanding necessary to carry out such experiments; nevertheless the long-term possibility of developing and applying such techniques under the cover of nature's irregularities presents a disquieting prospect.'[141]

Practical experiments to try out these ideas have either not been carried out so far, or have been kept secret by the military. If the track of tropical storms could be directed this capability could surely also be used for peaceful purposes in a profitable way. But the practical aspects facing the decision makers would be rather large and complicated.

Another kind of wind vortex, the tornado, occurs in the USA, rather more short lived and affecting smaller areas. Computer-simulated investigations of whether and how tornadoes can be influenced or prevented have not yet reached a sufficiently advanced stage to allow an assessment of the risk associated with practical experiments.[142] Commercial weather modifiers involved in hail prevention in Texas and Kenya have on several occasions seeded tornadoes which accidentally crossed their operational area. As a result changes have been observed in the radar echoes and a considerable enlargement of the precipitation area associated with the tornado, but no changes in the strength of the wind.[143]

Can a large-scale weather situation be modified?

Meteorological forecasts prepared by computer predict tomorrow's weather as well as can be expected under present circumstances; the quality of these forecasts is likely to improve somewhat when bigger computers become available, allowing a more detailed input of data. Continuous comparison of the predicted and the actual weather reveals mistakes and shortcomings in the mathematical model and enables it to be adjusted more precisely to the actual weather processes. Small discrepancies and inaccuracies in the forecast might thus be gradually diminished. However, there remain isolated occasions when forecasts turn out to be surprisingly bad. The Canadian meteorologist D. Davis investigated these unexpected mishaps and came to the conclusion that in the evolution of weather there are 'branching situations', when two different ways of development are equally likely, but any other development is improbable.[144]

A typical example is provided by the movement of a low pressure area from the south towards the southern tip of Greenland. It is going to move either eastwards in the direction of Iceland and the Norwegian Sea or northwestwards into the Labrador and Davis Straits. Occasionally the low will divide – one part moving east, the other northwest. Experience shows that low pressure systems never take the straight path across Greenland. Depending on the branch a depression follows, weather in Europe and North America will develop one way or the other and forecasts soon depart far from reality if the computer selected the wrong branch.

A branching situation may also arise when two low pressure areas develop relatively close to each other. Only one of them reaches full strength, the other soon disappears. If the wrong horse has been backed, the forecast will differ greatly from the actual development.

Davis identified seven such meteorological branching situations and assumes that further search will reveal more. Some of these appear to be indeterminate, both branches remaining open for a while. In others, the cumulative effect of unavoidable small inaccuracies in the initial data (boundary conditions) which are fed to the computer lead to bad forecasts. This suggests the existence of unstable conditions under which a small but well-targeted human intervention might switch the development from one branch to the other.

Joanne Simpson also thinks it possible that carefully targeted intervention at key points could produce large-scale effects.[145] She points out that depressions, especially in the Gulf of Alaska, sometimes intensify 'explosively'. This may be the start of a chain of reactions leading to changes both in the pattern of the hemispheric air circulation and also in the large-scale weather. A weak depression can presumably be intensified by seeding of its cloud systems, and thereby a change in the large-scale weather situation might be artificially induced.

Another key area for weather modification is the equatorial low pressure belt. Several thousand mighty cumulonimbus clouds, so-called 'hot towers' always found in this region, provide one of the most important energy links in the large-scale atmospheric circulation. If changes can be induced by precision seeding, recognisable effects on the weather may extend into subtropical regions.

There was a time, writes Dr Simpson, when Langmuir was ridiculed when he thought that cloud seeding experiments in New Mexico might influence the weather at Boston, more than 3000 km away. Today, we know enough about atmospheric processes that a wise meteorologist will neither assert, nor riducle, anything. Rather he will carefully examine and check any hypothesis, including those on extended area effects.

Low pressure areas over tropical oceans could be intensified by massive seeding with soot, according to the American geophysicists Gray and Frank. This would involve an operation much more elaborate than cloud seeding. Whole shiploads of fuel oil would have to be converted into soot in the modification area by specially constructed burners. Heat thereby generated could intensify the intended meteorological effect. Whether such a big effort is worthwhile, and whether it is compatible with environmental aspects, remain open questions. Nevertheless, the two scientists believe that the low pressure vortex created by precise soot seeding and the associated clouds could, after a day or two, bring precipitation to a neighbouring area which is suffering from drought. They also think it possible that low pressure areas in moderate latitudes, for instance in the drought-stricken western states of the USA, could be intensified by soot seedings.[146]

So far, the discussion has revolved around the creation, modification and intensification of *low pressure* areas and their associated cumulus cloud systems. In the Soviet Union, large-scale seeding has been carried out on stratus clouds and subsequent changes in weather resembling the development of a weak *high pressure* cell have been observed.[147] The seeded areas – several thousand square kilometres – remained cloud-free for several hours after the operation, while the alto-stratus deck over the neighbouring unseeded areas remained unbroken. As sunshine warmed the ground of the cloudfree region, temperatures rose 7–8 °C above those in the neighbouring overcast regions. This was sufficient to break up a temperature inversion temporarily. This dissolution of stratus over large areas which is possible with the presently available techniques could, according to Professor Federov, serve as trigger mechanism for changes in large-scale meteorological conditions. However, in recent years these experiments inspired by Professor Federov have apparently not been continued.

Only ideas and suggestions have been discussed in this section. As far as I know, their practical application has not been tried anywhere. However, they show that the concept of controlled changes of large-scale weather is no longer a Utopian idea but the subject of study by serious scientists. Apart from raising technical and scientific questions, there are numerous socio-economic, political and legal problems. These will be discussed in the third part of the book.

Ice ages and their causes

It is 'normal' for climate to change. The average values of weather elements for the last 30 years are different from those of previous 30-year periods; the averages for the next 30 years will presumably be different again. Four hundred years ago, glaciation of European mountains was much more extensive than today and some scientists call this period the 'little ice age'. A thousand years ago, when the Vikings came to Greenland they found areas covered by grass along the coast where only ice is seen today. The ice sheets then terminated in the interior.

If longer time periods are compared, the range of variation becomes even bigger. The 'normal' conditions of the last 300 000 years were ice ages. There were only short interruptions, the interglacials, during which climate was similar to that of today. Looking at the last 300 000 000 years, the 'normal' conditions meant less ice than today, and correspondingly higher sea levels. This warm 'normal' was only occasionally interrupted by relatively short ice ages.

Life in the oceans is more than 3000 million years old, and life on the continents over 400 million years. Since earthly creatures can exist in only a relatively narrow temperature range, it must be assumed that the climate of our planet has remained throughout this period within the limits in which life is possible. The variations of mean annual temperature have probably not exceeded 20 °C during the last 3000 million years. Compared with the seasonal changes of temperature in moderate latitudes, this is not impressive, but even small changes in the mean annual temperature are of great significance and can have incisive effects. The global annual mean temperature during the height of the last ice age was 'only' about 6 °C lower than today.

Superimposed on the large variations which appear to be associated with global warming and cooling are smaller fluctuations, often with different effects, in different regions of the globe. The equatorial low pressure belt moves north or south in association with shifts of the climatic zones so that a warming in moderate latitudes of the northern hemisphere corresponds to a cooling in the corresponding latitudes of the southern hemisphere and vice versa.[148] The drought in the Sahel zone occurred at the same time as rainfall increased over the northern parts of the Sahara. Also there are east–west movements of quasi-permanent pressure features, such as the shift of the 'Icelandic Low' in the direction of Greenland. Such shifts accompany general adjustments in the pattern of the global circulation, and of the prevailing winds in certain regions. As a result, the paths of warm and cold air and ocean currents are altered and the climate changes.[149]

There are many speculations on the causes of climatic fluctuations but, as yet, none of the theories proposed has been generally accepted. Most likely, many factors combine to influence the climate. Some of the factors proposed are changes in solar activity in response to processes in the sun's interior, passage of the solar system through cosmic dust regions (spiral arms of the galaxy), changes in the elements of the earth's orbital motion, continental drift, and changes in the transparency of the atmosphere by volcanic eruptions. The relative importance of each of these is still disputed.

For an explanation of climatic change over ultra-long periods (hundreds of millions of years), the effect of continental drift is certainly of great importance. At present, nine-tenths of glacier ice is in the Antarctic. If this ice were to melt, sea levels would rise by about 60 m. In earlier geological ages, when the tectonic plate that carries the Antarctic continent was still situated in middle latitudes, both the southern and the northern polar regions were covered by oceans. This implies that, quite apart from temperature variations, there could not have been as much ice as today and that sea levels were correspondingly higher. The drift of a continent into the south polar region allowed a massive build-up of ice and must have had enormous repercussions on the climate all over the globe.

The present distribution of continents shows a preponderance of land masses in the subpolar and middle latitudes of the northern hemisphere, while corresponding latitudes of the southern hemisphere are covered by oceans. The build-up of ice is therefore primarily a concern for the higher latitudes of the northern hemisphere. Glaciers grow when more snow falls than melts or evaporates. Therefore, one of the conditions likely to be conducive to the start of an ice age is a large amount of precipitation in not very cold winters combined with relatively cool temperatures in summers. This combination is periodically produced by the long-term changes in the inclination of the earth's axis, by the precession of the equinoxes and by the changes in ellipticity of the earth's orbit about the sun.

The idea that variations in astronomical elements of the earth's orbit are the main reason for ice ages appears in the 1875 edition of Sir John Herschel's *Outlines of Astronomy*. Among its strongest proponents were the British natural scientist J. Croll and the German meteorologist Alfred Wegener (posthumously famous for his theory of continental drift). The Yugoslav scientist M. Milankovitch subsequently calculated the problem in detail and developed a well-founded theory which has found more and more followers in recent years.[150] Changes in the orbital elements of the earth, and the resulting changes in the incident solar radiation in various latitudes, can be obtained very accurately with modern computers. Examination of drill cores from the sea bed allow deductions to be made about the climate of past ages. They show surprisingly good correlations with the changes predicted by the Milankovitch theory.[151]

If this theory is correct, then long term forecasts can be made. At present we appear to be at the end of an inter-glacial and would have

100

to assume that ice will grow considerably in the next hundreds and thousands of years – unless this evolution is counteracted by a human intervention which leads to warming. On the other hand if ice ages occur only during the passage of the solar system through a spiral compression lane in our galaxy – an argument put forward by W. H. McCrea, a British astronomer, and by G. E. Williams, an Australian geologist[152] – then we are leaving a spell of ice age behind us and are entering a warm epoch which will probably dominate the next 250 million years.

All these external factors, whether they come from the sun, from the planets or from galactic dust, influence the interactions between the numerous processes in the atmosphere and oceans of our own planet. Among the many interplaying processes are a number of feedback mechanisms with varying response times, which provide a certain stability to the system, while allowing it to vary within narrow limits. Attempts to model this exceedingly complex system mathematically are still in the initial stages and the results so far are only roughly approximate to the realities of nature.[153] Forecasts of future climatic changes must therefore be considered with a fair amount of scepticism.

The spreading of ice is assisted by its interaction with solar radiation. Newly formed ice and snow reflect sunshine more strongly than the land or ocean surface that was there before. As less energy is absorbed by the earth's surface, the air cools and the melting of ice and snow is counteracted. Volcanic activity is another factor that appears to be linked with ice ages. Drilling cores from the ocean floor show that an unusually high amount of volcanic ash must have been in the atmosphere at the beginning of the last glaciation. However, it has not yet been possible to establish cause and effect from this observation. It is conceivable that the growing glaciers increase the pressure on certain plates of the earth's surface thereby producing tensions in the crust which trigger the volcanic activity. Also, the increase in turbidity of the atmosphere caused by volcanic eruptions would reduce solar radiation at the ground and have a cooling effect. It would be that both these processes mutually amplify each other to produce the ice age.[154]

Not only positive (amplifying) but also negative (damping) feedback effects are built into the overall system. Lower temperatures and a reduction of the area covered by oceans due to spreading ice. causes evaporation, atmospheric water vapour content and precipitation to be reduced. Studies of prehistoric levels of African lakes show that 20000 years ago, at the height of the last glaciation, large parts of Africa had been much drier than today.[155] Less precipitation means a reduction in the supply of raw material for the growth of ice. Glaciers are, so to speak, starved by the atmosphere and eventually melt down. This, and possibly other feedbacks have prevented the ice from covering the earth altogether.

If the Milankovitch theory turns out to be right and if we are standing on the threshold to a new ice age, glaciation should start to increase on the mountains of the northern hemisphere and the seasonal spring melting should be delayed more and more. In this case, the climatic zones with their typical flora and fauna would shift equatorwards (and downwards in the mountains). This development would cause a marked climatic deterioration in the highly industrialised countries of middle latitudes and probably also in the Sahelian zone of northern Africa. In other regions, such as the Mediterranean, there could be a corresponding amelioration – at least during the first few hundred years.

Obviously, some people will gain and others will lose from a climatic change. This begs the question of whether such changes could be caused inadvertently, or even intentionally, by man's activities. Of course, man is not capable of tilting the earth, of changing the shape of its orbit or of varying any of the other external factors. However, he could conceivably influence the feedback and amplification mechanisms which retard or accelerate the development of a new ice age. Gordon MacDonald refers to a hypothesis of A. T. Wilson, according to which a sudden acceleration of the flow of Antarctic glaciers could increase albedo and lower the temperature enough to trigger an ice age.[156] If this hypothesis is correct, then such a development could be initiated by nuclear explosions at the base of the Antarctic ice cap.

More realistic than these rather speculative considerations, appears to be the anxiety about various human activities contributing to a warming of the atmosphere. This would shift the climatic zones towards the poles, especially in the northern hemisphere.[157] Benefits would accrue from this warming for the high latitude regions and,

102

presumably, for the Sahel region; but for large regions of the subtropics, such as California, the Mediterranean, North Africa, the Near and Middle East, such a shift might be catastrophic. Many scientists concerned with the possibilities of climate modification no longer attach the highest priority to the question of improving the present conditions, but rather to the problem of *avoiding* any natural or inadvertently caused climate variations.[158]

Climate modification – change or stabilisation?

Thousands of years ago, when man could only pray to his gods in the hope of changing the weather, he had already modified climate by his activities. Since the beginning of agriculture after the end of the last ice age, civilisation has been spreading at the cost of large amounts of forests. Most of the presently cultivated areas – fields, pastures, and built-up land – were once covered by forests. So were many areas which are barren today as a result of human action: kerst, barren mountains covered only by low growth, heaths, and even deserts. It is not always possible to be sure which changes have been caused by natural climatic fluctuations and which by human intervention, but there is no doubt that forest clearance and overgrazing have intensified desertification and enlarged arid regions.

Forests influence the environment in a variety of ways. They break the winds and diminish the amount of dust which is carried up by them, polluting the atmosphere. They filter out dust and fog particles from the air. They act as large sponges, retaining precipitation and passing it slowly on to the rivers. Their rate of evaporation exceeds that of agricultural land and arid land greatly; they thereby contribute significantly to the moisture content of the air. From the atmosphere they take up carbon dioxide and fix it for prolonged periods at the same time giving off oxygen. All this suggests that the reduction of forests by man's intervention must have affected the climate: it could have contributed to an increasing frequency of droughts in eastern Europe and also caused a gradual diminution of rainfall in the Mediterranean region.[159]

The urbanisation and industrialisation of the present age which, presumably, will continue to increase further in future, have an effect on climate and environment. Towns and large industrial complexes are 'heat islands' with temperatures several degrees above those of the rural environment. Other meteorological elements also vary between town and country. American studies based on average values from several cities in comparison with open country showed that in urban areas visibility can be reduced by 26%, wind velocity by 25%, sunshine by 22%, and air humidity by 6%, while the incidence of fog can increase by 60%. These effects extend downwind to distances of up to 50 km and more, where increases in cloud cover of 8%, in precipitation of 14% and in thunderstorms of 16% have been found.[160]

The hydrological cycle has also been considerably influenced by man. Reservoirs, irrigation of agricultural land, cooling towers and industrial utilisation add to evaporation, thereby somewhat counteracting the reduction in evaporation by deforestation. It is not quite clear whether all these effects balance in the end, but it is certain that local redistributions of the humidity in the atmosphere occur.

Urbanisation and industrialisation not only produce local changes, but may also influence weather on a world-wide scale. Rapid increases in burning of fossil fuels and massive deforestation, especially in the tropics, have led to an increase of the carbon dioxide content of the atmosphere of 15% (from 290 to 335 parts per million) since the beginning of this century and a further rise of 20% by the year 2000 is predicted. Radiation from the sun passes unhindered through the carbon dioxide in the atmosphere to the ground. But the heat radiation from the earth to outer space is absorbed by the carbon dioxide and radiated back. This is the 'greenhouse effect' which is thought to lead to global warming in response to carbon dioxide increases. The discovery that the biosphere is not a sink for the carbon dioxide but rather a source (because of deforestation) is a rather new problem for climatologists and forces them to revise thoroughly their existing model concept of the carbon dioxide cycle.[161] Estimates of the size of this source differ considerably at present. However, there are good reasons for assuming that the oceans have taken up considerably more carbon dioxide in the past than meteorologists and oceanographers have estimated and that they will probably be able to take up these amounts in future. Forecasts concerning the future evolution of the carbon dioxide content of the atmosphere are fraught with uncertainty. It is likely the situation is not as dramatic as was feared a few years ago. But even with optimistic assumptions, the conclusion is that carbon dioxide increases expected during the next 100 years could have far-reaching effects on the climate.[162]

This is the more important, as other human activities could also contribute to warming. There are other gases besides carbon dioxide which as a result of human activities increase the 'greenhouse effect'. Among them are nitrous oxide, which is produced by decomposition of fertiliser, and the 'freons', which are used as propellants in aerosol cans. More and more waste heat is put into the environment with increasing energy consumption. The effect of dust particles from agricultural and industrial activities on the temperature is difficult to assess; it depends on their vertical distribution in the atmosphere. Dust reduces the radiation transparency of the atmosphere and reflects and absorbs part of the incoming sunshine, preventing it from reaching the earth's surface. The air is warmed by the absorption of sunshine at the levels where the dust is concentrated. According to the view of the German climatologist Hermann Flohn, the combined effects of all these factors could produce a northward shift of the climatic zones by several hundred kilometres. This would have disastrous effects for the agriculture of the third world.[163]

The Soviet climatologist M. I. Budyko, who commands great respect in western countries, points out that an assessment of the economic effects of a general warming must not be based only on the adverse effects on agriculture in middle and subtropical latitudes.

Undoubtedly the warming will have not only negative but also favourable effects upon many branches of national economy. For example, a longer vegetation period would be important for agriculture. Warming can essentially improve navigation in the Arctic latitudes and facilitate the exploration of the polar regions. However, considering these and other favourable consequences of the climate warming, it has become clear that an essential change in global climate would not be desirable because national economies of different countries of the world are adapted to the present climate.[164]

Budyko and his collaborators therefore investigated ways and means of combatting a general warming.

Weather modification can be used not only to combat climatic changes inadvertently produced by man, it can also serve to improve

105

natural climates. Professor Federov thinks it unlikely that this can be put into practice during the next two or three decades. Large-scale changes in the climate can only be brought about by interventions which 'set off a chain of events' in a determined direction and thereby induce the required adjustments to the general circulation pattern of the atmosphere. To be effective, the pattern must be persistent and it should be predicted without error; here, not in matters of technology or expenditures of energy, lies the fundamental and staggering difficulty of the problem.'[165]

Besides the local or regional scale projects, there is the possibility of interventions that have much larger consequences, such as a deflection of the cold and warm sea currents which determine climate. A favourite subject of speculation regarding large-scale climate modification is the Arctic Ocean. There, one can visualise the possibility of a 'switch' to another relatively stable situation. At present, the Arctic is covered by ice, 2–3 m thick, which reflects solar radiation strongly and keeps the region cold. Even in summer, much of the ice cover never melts. If it were possible to remove the ice at any time, the reflective property of the Arctic would be drastically reduced; this could lead to warming and, possibly, to a change in the atmospheric circulation in the sense that the arctic basin would probably remain free of ice for a long period.[166]

Some rather unrealistic proposals have been put forward, regarding the techniques that could be applied:[167] blasting of the ice cover with hydrogen bombs which, at the same time, would bring up warmer water from the depths to the surface (associated with serious damage to the environment); spraying of soot onto the ice (enormous amounts of material would have to be transported); erection of a dam, with turbines in the Bering Strait which pump cold Arctic waters into the Pacific so that more warm Atlantic water gets into the Arctic basin (considerable infringement on the environment of the Pacific, large investment costs, continuous large energy consumption). Smaller projects would probably suffice to produce a similar effect. The Arctic Ocean has a relatively thin 'lid' of cold, low-salinity water, 60% of which comes from the Siberian rivers and 40% from the water flowing through the Bering Strait. Underneath is relatively heavy, high-salinity water from the Gulf Stream. According to a long-standing project, the Soviets intend to divert part of the waters

106

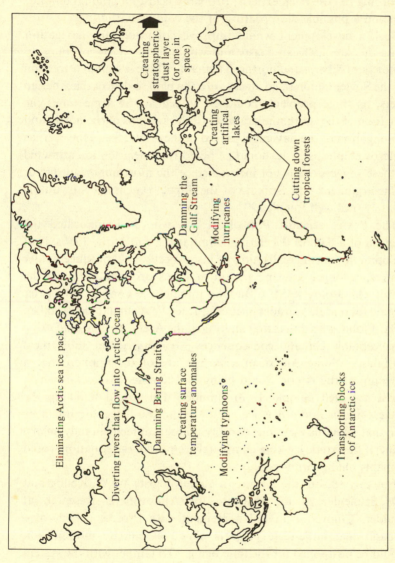

Fig. 19. Various schemes of climate modification that can be achieved with present-day technology. (After W. W. Kellogg and S. H. Schneider, *Science*, **186** (1974), p. 1163.)

Creating stratospheric dust layer (or one in space)

Creating artificial lakes

Cutting down tropical forests

Damming the Gulf Stream

Modifying hurricanes

Eliminating Arctic sea ice pack

Diverting rivers that flow into Arctic Ocean

Damming Bering Strait

Creating surface temperature anomalies

Modifying typhoons

Transporting blocks of Antarctic ice

of the Pechora, Ob and Yenisei rivers into the drier regions in the southern Soviet Union. The realisation of this project would, according to American experts, probably lead to a gradual diminution and a possible disappearing of the cold water lid.[168]

Such a development would improve conditions for navigation along the Arctic Ocean coasts and would facilitate the opening up of northern Siberia and northern Canada. There is a great attraction for the Soviet Union to go ahead with the plan to deflect the Siberian rivers, as three problems could be solved at the same time; the irrigation of the south, the drainage of the swamps of the north, and prolongation of the navigable period in the Arctic Sea. However, as Federov emphasises, 'it should be noted that the attention attracted to these suggestions is not merited since the most important factor in the argument – an estimate of the impact on climate – is, unfortunately, not well founded.'[169]

The Soviet scientist P. M. Borissov, who had advocated the erection of a dam in the Bering Strait at the end of the 1950s, is of the opinion that the realisation of his project would produce milder winters, a longer growing period and less droughts,[170] while the Soviet climatologist M. I. Budyko predicts a considerably drier climate for middle latitudes in consequence.[171] A similar view is taken by H. Flohn, who thinks that melting of the Arctic ice lid would bring 'unacceptable climatic consequences' to countries in subtropical latitudes.[172] Some American scientists are of the opinion that a warming of the Arctic would bring an increase in snowfall which might not melt completely in summer. This could build up the glaciers and herald a new ice age.[173] From the foregoing it can be seen that opinions vary considerably. There is agreement only on one point; if the Arctic ice were completely removed, the situation would probably change irreversibly.

For this reason, the American climatologists W. W. Kellog and S. H. Schneider of the Institute for Atmospheric Research at Boulder, Colorado – in full agreement with Professor Federov – demand that large-scale projects aimed at climate modification should be postponed for the time being. 'To tamper with the system which determines the livelihood and lifestyles of people the world over would be the height of irresponsibility if we could not adequately foresee the outcome.'[174]

3
Problems and dangers

Should the weather be modified at all?

In spring 1972, prior to the start of a large-scale experiment on hail prevention and rainfall stimulation in South Dakota – which has since been abandoned – the authorities sounded public reaction to weather modification and later repeated such enquiries throughout the duration of the programme each autumn.[1] These opinion polls were supposed to clarify whether and how public opinion was changing as a result of practical experience obtained during the weather modification operations. One of the questions asked was 'do you think that cloud seeding probably violates God's plans for man and the weather?' The percentage distribution of the answers to this question which was asked at the beginning of the project and again after two seeding seasons in autumn 1973 is as follows:

	Spring 1972	Autumn 1973
Yes	43	38
Uncertain	14	15
No	43	47

The question 'Do you believe that cloud seeding programmes, even if carefully controlled, are very likely to disturb the balance of nature?' was answered:

	Spring 1972	Autumn 1973
Yes	42	40
Uncertain	21	23
No	37	37

A question posed earlier as to whether a cloud seeding programme should be initiated if there was the prospect that farmers would derive help from it was answered as follows:

	Spring 1972	Autumn 1973
Yes	69	72
Uncertain	11	8
No	20	20

Similar opinion polls in other states produced similar results.[2]

The relatively high number of 'yes' answers to the two first listed questions is in contradiction with the considerably smaller number of 'no' answers to the third which was chronologically the first of a large number of questions put by the interviewers. The American sociologist J. Eugene Haas explains that weather modification was of no real concern to most of the people interviewed and they had not given the matter much thought; therefore they had no strong opinion on the problem and did not notice when in the course of answering many questions they gave partly contradictory statements.[3] The frequently encountered religious aversion against weather modification could not be traced to any official doctrine of the known religions.

There is no scientific basis for discussing religious attitudes. The argument that weather modification unbalances nature is based on the assumption that science will never completely unravel the complicated and numerous interrelationships of nature or be able exactly to predict the consequences of human intervention over prolonged periods. There is also the risk that human beings become addicted to an artificially created environment. In trying to solve a particular problem we might well resort to weather modification; but, since evolution is never quite as expected and man is unable to dominate nature perpetually, even larger problems might be created in the long run.[4]

If presented in this way, without the usual exaggerations, this argument must be taken seriously. But the lingering desire to return to a natural environment can unfortunately no longer be fulfilled. The 'back to nature' path was blocked when our ancestors began to cultivate the land about 10000 years ago without a 'technology assessment' – an estimation of the long-term consequences of this venture. The final result is what we call 'human civilisation' – with all its advantages and problems.

When man began to create elements of an artificial environment by growing cereals and other 'artificial' vegetation, their obliteration by natural plants (weeds) had to be prevented by human intervention. This started a chain reaction wherein the solution of a problem led to the creation of another larger one. A growing food supply allows the population to increase and creates the need to augment the food supply further. In this way, mankind took the long road from the beginnings of cultivation to present-day irrigation and fertilisation of agricultural land step by step through an ever-increasing 'artificial' environment. During the course of this evolution, the population has grown by a factor of 1000 during the last 10000 years. Today we realise with horror that this progression has brought us to a limit of growth which cannot be surpassed.[5] We face the danger that a continuing population explosion might in the end destroy our environment, the very basis of life on earth.

It is questionable whether a return to a 'natural environment', in which man does not attempt to dominate nature but rather is governed by it, is really desirable. In any case, such a voluntary return is no longer feasible, as the present world population of thousands of millions of people could not live off the earth under 'natural conditions'.

In this rather problematic situation, we must look for a solution which does not increase the strain on the already stressed environment. We may not be able to do without 'artificial environments', but we must abandon the idea of 'dominating' nature regardless of the consequences. We must try, as much as is possible under the circumstances, to co-operate with nature,[6] and the better its laws are understood, the more chance there will be of success. While human understanding of nature will always be incomplete, we must at least apply sensibly whatever understanding we have. Unfortunately this is rarely the case today.

The solution to the immense problem of creating an acceptable and dignified lifestyle for a rapidly growing population, without destroying the environment, does not require that we should abandon our achievements in science and technology, but rather that we apply them according to careful worldwide planning.[7] In this context also, weather modification can be useful, but only then if a thorough assessment of the technology is carried out, so that at least the undesirable consequences, which can be predicted with our present knowledge of relationships between nature and society, can be avoided.

At long last, representatives of scientific disciplines outside meteorology are beginning to ponder such problems. This is by no means premature, although many weather modification techniques are still in the development stage. Had these discussions started during Langmuir's time, instead of arguing about weather modification being at all possible, we would perhaps have found a better alternative to the financing by the military and by private enterprise of a development with such far-reaching consequences.

According to Canadian geographer W. R. D. Sewell,

> the assessment of weather modification had so far concentrated on the question of feasibility and the discussions were led by two prominent groups: scientists who are principally interested in understanding the atmosphere and who are therefore motivated towards obtaining further financial support for this purpose, and representatives of mission-oriented agencies who see in weather modification a possible means of pursuing various possibilities and possibly increasing the social and political importance of their agency. Understandably too, the evaluations of the second group have been colored with some enthusiasm. *Unfortunately it is still not clear whether weather modification is socially desirable and, if so, in which situations.*[8]

Sewell, like any scientist, is interested in obtaining more funds for the branch of science in which he is interested. But he does *not* present a narrow, personal view, biased by specific interest, when he complains that of the 1973 budget voted for weather modification in the United States 98.5% were earmarked for physical research and technical development, and only 1.5% for the investigation of the human and ecological aspects. Even more disquieting is the hint that apparently not enough is done to interest representatives of other disciplines in this problem, and that not even the envisaged 1.5% had been fully utilised.[9]

Sewell's question on the social desirability of weather modification is theoretically justified but, in practice, already overtaken by events.

112

The development has gone too far to be abandoned altogether; neither is it possible that current operations be stopped or even suspended until it has been settled whether the possible disadvantages for mankind are, perhaps, much larger than the expected advantages. The possibility that negative consequences could arise in the long run has in the past never led to the termination of a technological development of which its proponents expected gain, scientific standing, or 'interesting' defense applications. 'Policies are discussed as possible means of use of weather modification, based on the presumption that desirable results will follow – and with little consideration of possible ecologic disadvantages', says M. Holden Jr, professor of Political Science at the University of Wisconsin.[10]

But with so much at stake for ourselves and future generations, it is not too late to ask when? where? how? and what for? The direction of further development of weather modification is probably still amenable to correction if special danger zones can be recognised at an early stage.

The problems which arise in this connection can be divided into two major groups: undesirable side effects and diverging weather interests. Among the first are all those consequences which cannot be foreseen by an individual weather modification expert, even with the most sincere professional approach. This includes extended area effects and possible consequences for public health, environment, economy and society. For example, to assess how a planned 20% increase of summer rainfall would affect the distribution of agricultural pests, the frequency and intensity of diseases of the respiratory tract, or the expected revenue from tourism in the area, is a difficult task for experts from various disciplines. Opinions are likely to diverge, but these questions *can* be discussed on a scientific basis. Such discussions can lead to the conclusions that, 'according to the view of experts', certain consequences might arise with a given probability. The results can be checked against such predictions and expectations and thus contribute to experience and allow corrections to be made in the future.

But such probability forecasts are only part of a thorough technology assessment. *Evaluation* of the possible consequences of a new technique cannot be objective in the scientific sense since it relies on a basically subjective scale of values. We cannot make an 'objective' comparison of the advantage of an envisaged 10% increase in the yield of barley versus the disadvantage of a probable 3% increase in the frequency of a certain respiratory disease. Matters are even more complicated when one deals with 'values' which cannot be expressed numerically. How can we quantify the consequences of weather modification that might cause the extinction of a certain animal or plant species in an area? Who is in a position to make relevant decisions?

These questions lead into another sphere of problems which arise because of people's different interests, not only with respect to unpredictable side effects and consequences, but even about the weather itself. Although it may not be difficult to agree on the idea that weather modification should serve a common good and not the interests of small but influential groups, how is 'common good' to be defined? Opinions diverge on this question and the problem cannot be discussed fully in a book such as this. It must suffice to point out a few concrete problems which will arise automatically once weather can be influenced by man and is no longer a natural phenomenon. Such problems should be dealt with before they become really pressing.

Small efforts – bad effects

Weather modification is relatively new. Occasional mishaps do occur and an operation may sometimes produce the opposite of what has been desired. For instance, before it was known that cloud seeding increases precipitation only within a certain temperature range and that at sufficiently low temperatures it can even lead to a reduction in precipitation, some commercial weather makers occasionally may have done their customers a disservice. Even today, rain making in cumulus clouds or frontal disturbances sometime might cause unpleasant surprises. The risk of reducing precipitation by improper seeding is probably of the same order of magnitude as the gains. It is hardly likely that modification could accidentally trigger a long period of drought which would otherwise not have arisen; but, in some sensitive regions of the world, a reduction of precipitation by only a small percentage would have rather serious consequences.

Similar problems arise in hail protection, with an even greater risk of damage if the opposite of what was intended occurs. This is also the possibility of seeding operations influencing the frequency and intensity of lightning strokes.

Besides the risk of producing the reverse of what is desired, there is also the possibility of weather modification operations being too 'successful'. Rain makers have been accused of triggering flood catastrophes or at least been held partly responsible for them. Hitherto it has however not been possible to prove such connections, which, in the opinion of meteorologists, are improbable.

The best known case is the break of the dam in Yuba City, California, after torrential rain at Christmas 1955. The floods killed more than 60 people and damaged property to the extent of $200 million. About 150 people who had suffered damage went to court and demanded compensation from the State of California, the authority responsible for the maintenance of dams. When the lawyer representing the plaintiffs heard that the Pacific Gas and Electric Company had undertaken orographic cloud seeding operations before the dam broke, he extended the claims for damage to this Company and to the weather modifiers employed by them. During the course of the proceedings, representatives of the public utility succeeded in convincing the court that the larger part of the supplementary precipitation produced by their seeding had fallen on the catchment of a reservoir further upstream whose dam had not overflowed during that period. As the seeding increased the water level only in the upper reservoir it could not have contributed to the catastrophe. Besides, the rain makers had stopped their operations during the period of heavy rains. In the end, the electricity company was cleared of guilt in a process which lasted, including preparation time, for more than ten years.[11] D. E. Mann, a political analyst, is of the opinion that the case would have ended differently had there not been the second defendant, the State of California, which is responsible for the maintenance of dams. Mann argues that it would have been very difficult to avoid giving any compensation to 150 heavily damaged families. As the case went on, the lawyers representing the two sides made a deal: the plaintiffs would not appeal against a dismissal of the case against the public utility and the weather makers; in return, the attorney of the company would refrain from calling more witnesses, or to instigate any measures that would prolong the process even further.[12] Although this case was finally decided in favour of Pacific Gas and Electric Company, the catastrophe of Yuba City was a bad blow for commercial weather modification. A number of public utilities subsequently stopped their seeding programmes and the turnover of the weather modification company involved in the Yuba City case was reduced temporarily to one-third.[13]

After the start of a seeding programme organised by the electricity company 'Quebec Hydro' in Canada, heavy rainfall set in and lasted for three weeks. Floods and landslides followed, inflicting heavy losses on agriculture and the tourist industry. The local population pinned responsibility on the weather makers. Despite the fact that all seeding operations had been cancelled during the period of heavy rainfall, 60000 people signed a petition which called for a complete ban on weather modification. The operators were informed that if necessary their activities would be stopped by force. The Minister for Natural Resources, René Levescue was threatened with murder if he did nothing to stop the rain. Even a report published by the government, explaining that the heavy rainfall had natural causes and was not produced as a consequence of the cloud seeding, could not pacify the masses.[14]

The reaction of people in South Dakota was quite different when Rapid City, situated in a cloud-seeding area, was hit by floods in June 1972, as officials had made a point of informing the population about the cloud-seeding programme. Subsequent public opinion polls showed that the majority of the town's inhabitants did not associate the flood with the weather modification activities. Even after the catastrophe, there was still a majority of those who endorsed the seeding programme.[15]

Little is known about the possibility of inadvertent extended area effects of weather modification. There are, as yet, no answers to the question of whether more rain in one place must lead to less rain in another. Also, it is not clear whether and under what circumstances it is possible to increase precipitation in a large region as well as in a relatively small target area. There are situations in which a mere redistribution of precipitation – for instance from the sea to the land – could be beneficial. It is not clear yet whether all cloud seeding is

116

only a mere redistribution of the precipitation. Of course, seeding does not create any additional water. Nevertheless precipitation produced by cloud seeding is not necessarily being 'stolen' from somebody else. Dynamic cloud seeding, by increasing the intake of moist air by clouds, might increase evaporation; and mutual interactions between degree of cloudiness, precipitation, rate of evaporation and changes in reflective ability as a result of snowfall are so complicated that there is no simple rule of thumb.

Systematic investigations into the possible remote effects of seeding were only recently included in some experimental programmes (Project HIPLEX, NHRE),[16] and significant results are not yet available. Post-mortem analyses of earlier experimental programmes can be based only on data from the permanent station network of the weather services which, for this purpose, is often not dense enough. However, such analyses show to the surprise of the meteorologists that orographic cloud seeding can produce increases in precipitation by 20–100% at distances of 100–200 km downwind of the target area. The probability that these are mere random effects is very small. Detailed analysis also shows that these remote effects always occur, or are particularly strong, when meteorological conditions (wind direction, cloud temperatures, etc.) are favourable.[17] Remote effects of the Colorado Basin Project appeared to extend to the neighbouring Rio Grande basin.[18] This was probably caused by the movement of seeding material (as actually observed in some cases)[19] and by the drift of the ice crystals which form as a result of the seeding. The resulting artificial cirrus clouds can produce effects in regions far from the target area (remote effects).[20]

Statisticians talk of 'positive' remote effects when the distant effect is of the same kind as the one in the target area, for example: increases of precipitation in both areas. It is likely that such remote effects would in most cases be felt as 'positive' by the people concerned. But what if the region affected by remote effects is threatened by avalanches? Or if abundant natural precipitation has already fallen and if further triggering of rain would produce flooding? Will the regulations which apply to the termination of seeding actions have to be extended beyond the target area, to take account of possible remote effects? Would weather modifiers or their clients compensate road authorities or municipal councils for damage

117

and increased snow clearing expenses in a distant area? Should possible benefactors of extended area effects make payments? Numerous questions arise which are difficult enough to answer if all this occurs within one country; but what if the remote effects extended across national boundaries?

Some 'positive' remote effects of Israel's orographic seeding programmes have been recorded at weather stations in the neighbouring Arab states (Lebanon, Syria, Jordan).[21] In this case, increased rainfall is welcomed by the surrounding countries. But could the weather modifiers in Israel know before the start of their programme whether the effects in the neighbouring countries were going to be positive and not negative? Would they have stopped the programme which is vital for their country if the remote effects on neighbouring countries were not desirable? Here there are questions involving international law which will be discussed more thoroughly later on.

Statistical post-mortem analyses on cumulus cloud experiments have not shown any significant trend,[22] but this is not surprising in view of the difficulties encountered in working with this cloud type. Positive and negative remote effects have been identified, and they are of an order of magnitude which is unlikely to have arisen randomly. Theoretically there could be several mechanisms which might trigger such effects.[23] The only thing we can be sure of at present is that remote effects of cumulus seeding vary strongly from case to case with the large-scale weather situation and with local circumstances, and will be difficult to predict. Inadequate knowledge also makes it impossible to foresee the mutual interaction of two or more seeding projects which take place independently of each other in neighbouring areas. Also, it is not clear which consequences would arise if one were to move from the present relatively small-scale operations to large-scale projects. But as long as these questions remain unanswered, it will not be possible to assess the economic benefits of cloud seeding for a whole nation.

Incorrect cost–benefit calculations

J. Eugene Haas writes:

Political and economic considerations usually blend together in the public decision-making process. Members of Congress and Federal Agency officials advocating the expenditures of tax revenues for weather modification frequently use economic arguments to support their favorite program proposals. The benefit–cost numbers game has become a standard part of the political process. Every agency has its favorite set of numbers purporting to show that the *potential* benefits from a proposed weather modification program greatly exceed the cost thereof. The fact that these numbers represent estimates usually put together by noneconomists without benefit of adequate data has not been seriously challenged in the literature. Indeed in one case where a group of economists working under Government contact reportedly concluded that the benefit–cost ratio of a proposed weather modification program was approximately one to one, their report was sent back to them for further consideration. [*This refers to the pessimistic assessment of the Colorado Basin rainfall project*.] This is not to say that the benefit–cost ratios may not turn out to be very favorable, but rather that frequently political considerations appear to have taken precedence over scholarship in the computing of the ratios.[24]

To the meteorologist, who knows more about the complicated mutual relationships and feedback mechanisms in the atmosphere than of those in world economics, it appears 'obvious' that in a time of threatening world-wide food shortage, weather modification could contribute towards an increase of agricultural products and thereby mitigate the hunger in the world. However, to the economist the relationships are not that simple. This world is full of contradictions. There are millions of people starving and hundreds of millions suffering from malnutrition. Yet, at the same time, there are 'butter mountains', 'milk floods', stores full of agricultural surplus products which cannot be marketed. The demand on world food markets is considerably smaller than the actual nutritional need.

Opinions about how this horrible contradiction could be resolved differ even more than those about the chances of successful weather modification. A discussion of this problem would go beyond the frame of this book. In judging the benefit of weather modification under *present* conditions, we must take cognizance of the world as it is, not as it ought to be. In the real world it makes little sense to increase the production of goods which cannot be sold. Nor does it make sense in cost–benefit analyses to quote prices for agricultural products which are artificially subsidised by the state.[25]

For the individual farmer, weather modification can mean protection from ruin by hail or drought. At the same time, a general 'overprotection' from the vagaries of the weather could lead to an oversupply of easily perishable products. The resultant price drop

119

would help the consumer but could inflict heavy damage on the producer who has to pay the weather modifier.[26] Problems of this kind exist in agriculture even without weather modification. They are the reasons for the existence of divers forms of state interventions and guidance even in those countries which profess a 'free market economy'. A realistic assessment of the possible benefits of weather modification to agriculture must therefore take into account how the application of this new technique fits into the complicated economic structure.

In principle, there are three sources of error which can arise in cost–benefit analysis. First, neglect of the fact that many weather modification operations involve not only a benefactor but also people who suffer damage; secondly, assessment based on too narrow regional perspective without adequate consideration of the effects on the total economy of a country; thirdly, insufficient attention given to long-term socio-economic consequences.

Those in agriculture who suffer from weather modification may be individual farmers who produce various types of crops and whose weather interests differ from those of the clients of the weather modifier. Artificial increases of rain can also damage the building industry (houses as well as roads), tourism (not to speak of the interests of holiday-makers which cannot be measured financially) and in many cases also other economic branches such as communications.[27]

There has been a backing of restricted regional interests in the USA especially by senators from drought-afflicted states who wanted to impress the electorate, but also by some official agencies who look after the interests of these states. Seen from an economic viewpoint, it is contradictory to assist agriculture in poor agricultural regions by artificial means, while the government must at the same time subsidise farmers in good agricultural land areas for *not* cultivating part of their land to prevent surplus production.[28] In the report of an interdisciplinary study, in which the sociologists J. E. Haas and B. Farhar, the meteorologist S. A. Changnon Jr, the lawyer R. J. Davis and the economist E. Swanson participated, it says:

Farmers as a whole will *not* gain from hail suppression technologies. Regionally and locally, some individual farmers will of course. But nationally, farmers will lose . . . because the price per unit product will fall. (Recall that past policies of land retirement helped restrict production and thus keep prices high.)...For the nation as a whole, improved hail suppression technologies are not particularly attractive economically...A major and perhaps the only significant national justification for large-scale application of an improved hail suppression technology is the use of agricultural products as an important component 'bargaining chip' in the U.S. foreign policy. If farm crops were to be used in much the same manner as military hardware is now used in the U.S. foreign policy, the increased crop production from hail suppression might take on a very special meaning and attract bipartisan Federal support. Such an eventuality may become more likely as the world population continues to outstrip world food production.[29]

Scientists correctly point out that cost–benefit ratios for agricultural weather modification are sensible only if comparisons are made with other possible alternatives.[30] Not only are there other more expensive but reliable methods of water supply, but we must also consider whether it would not be economically more prudent to reduce water consumption by growing agricultural crops which can thrive in a natural environment, and, if necessary, to breed them. Far-sighted economists, sociologists and ecologists put further questions: is it at all sensible to improve the water supply to the western USA by artificial means and thus to attract more people? Would it not be better to curb the influx of people into these regions that are poorly endowed with water by nature?[31] Is it really desirable to counteract the drift of young people into towns by improving conditions in agriculture? Should we not rather simply accept this drift as being a natural adaptation to the changing economic conditions?[32] Would it not be ecologically as well as economically more sensible to reconvert part of the agricultural land into forest? This might be of advantage to the water budget and the fertility of the whole region – inclusive of agricultural areas at lower elevations – and would, for some time to come, produce higher yields for the total economy of the country.

The way in which the west of the USA has been opened up is queried by some experts and the question is raised of whether weather modification in this area is desirable. Charles Cooper, professor for Environmental Research at the California State University, is of the opinion that the initial employment of weather modification experts in the prevention of catastrophes (droughts, hail) is a wedge in the door. If they are successful in doing that, they will then be employed in the augmentation of rain during 'normal' years. Agricultural production would then adjust to the newly created artificial weather conditions. Instead to producing the crops which grow best under natural conditions, there might be a switch to products which promise more profit. Pastures might be converted to fields and the danger of wind erosion would be increased. An increase of livestock would heighten the danger of overgrazing. Prices for land would rise and so would the profits of farmers. But, as the activities of weather modifiers can never be 100% reliable, now and then, a drought or hail damage would still occur despite their efforts. We must therefore take into account that, after a few decades, the damage caused by drought and hail would probably be larger than before the introduction of large-scale weather modification. This, because more profitable crops would be more sensitive to climate and the proportional amount of damage would consequently increase.[33]

These reservations and considerations are not to be understood to mean that the cited critics decline weather modification for agriculture under all circumstances. Rather, they reject the naïve optimism – or the optimism artificially created by interested persons – by which only the advantages of the new technique and not the problems connected with it are emphasised.

Some nonagricultural applications of weather modification appear to be economically more important, and less problematic. In these – as far as we can see – there are only beneficiaries and nobody suffers. This applies to the dissipation of fog at airports (if we exclude the loss of business of those who profit from deviations of aircraft) and to projects in forestry aimed at the prevention and combating of forest fires.[34]

Some negative effects (especially additional costs for snow clearing) accompany orographic cloud seeding operations aimed at augmenting the water supply to reservoirs of hydroelectric power stations. However, even after allowing for these effects, there remains a considerable economic benefit to the electricity-producing companies. Seen from a higher economic level under the present energy situation, any increase in hydroelectric power production that means a lesser import of fossil fuels is desirable. Furthermore, this kind of energy production does little damage to the environment.

122

Social and human aspects of weather modification

In the mid 1960s, the National Science Foundation of the USA established a small – much too small – group of nonmeteorologists to study 'human aspects' of weather modification. Sociologists, ecologists, economic and legal experts have done valuable studies in this framework and have revealed connections that would probably not have been seen by laymen and meteorologists. A survey of this work, passages of which were quoted in previous chapters, shows that the possible effects of weather modification have a bearing on many diverse facets of natural and social life: the development of property prices in agricultural areas of the Midwest, the population of elk in the Rocky Mountains, the building industry, sport fisheries and many other matters of importance to bigger or smaller groups in society. But basically it all concerns money. Even the elk is of economic interest, as some places in the Rocky Mountains owe their prosperity chiefly to the income from hunting facilities which they offer to rich clients.[35]

Nobody would deny that such studies are useful and necessary and that it is only reasonable to investigate whose business interests might be affected by weather modification. But are man's interests only of an economic nature? Are there no other human aspects which should be considered?

It was shocking for me to realise that in all the source material I read before writing this book, I never came across any reference to the effects of weather modification on the physical and psychological health of man. Bergeron, in his pioneering paper of 1949, points out that orographic cloud seeding might perhaps trigger föhn, but this possibility has apparently never been investigated. When public outrage in Quebec accused weather modifiers – probably unjustly – of having produced a long-lasting rainy period, some doctors maintained that the health of children and old people had been affected by the reduction in sunshine.[36] It is no longer possible to find out whether such statements were just manifestations of mass hysteria, whether unjustified generalisations were made from isolated cases, or whether a real problem did exist. Relevant statistical studies had not been undertaken. The same applies to other regions where weather modification was not followed by public outrage.

In the literature about human aspects of weather modification there is sometimes a general hint that weather influences human productivity, suicide and accident rates, and the frequency and intensity of certain illnesses.[37] However, no detailed studies of the positive and negative effects of weather modification seem to have been made by members of the medical profession. Neither are relevant medical studies mentioned in a specially prepared list of urgently required future research, which was presented in 1972 at a symposium at the National Center of Atmospheric Research in Boulder.[38] Professor W. R. D. Sewell, the director of the symposium, in reply to my enquiry, said that he was not aware of any major investigations into health aspects of weather modification.

A detailed analysis of possible ecological consequences of weather modification by Charles F. Cooper and William C. Jolly contained the statement that direct effects on man are not included in the subjects for investigation.[39] However they point out the possibility that the spread of pathological vectors, such as mosquitoes, could be increased by additional rainfall.[40] Weather modification could also affect mammals that transmit fatal human diseases such as plague, rabies, yellow fever, etc. Both scientists are of the opinion that this would not cause any major problem in the USA, but that a large-scale application of weather modification in the tropics should be preceded by thorough investigations in this respect.[41]

The neglect of human aspects in weather modification studies is shown not only by the inadequate consideration of medical questions but also by the actual motivation of the projects. All applications so far served either military or economic purposes. Fog is dispersed to serve commercial or military aviation interests; additional precipitation helps power stations and agriculture, or assists the military to hinder the supply of enemy front lines, lightning is prevented to protect forests and so on. Nowhere have I found any consideration as to whether making fair weather for man as a human being (rather than as an economist, consumer or soldier) might be considered.

Notice has been taken of the interests of tomato growers in southern Florida, who need dry weather in April and May, by abstaining from cloud seeding operations during these months.[42] But a suggestion by Crutchfield[43] that seeding operations should not be done on weekends in the interest of recreation has, as far as I know, not been followed up. If such suggestions are even taken seriously in future, it will probably be in the interest of tourist enterprises rather than to satisfy the stressed members of the society whose longing for a fair weekend cannot be measured in monetary terms.

Of course, the technical capability to produce fair weather on demand is still rather limited. The present state of weather modification does not yet allow the passage of fronts or the movement of low pressure areas to be influenced. But at least one kind of 'fair weather' production is theoretically possible already: the dissolution of supercooled fog and stratus cloud, not only above airfields but over larger regions – in the interests of people who like sunshine. In autumn and winter prolonged high pressure situations occur in middle latitudes. Often there is brilliant sunshine on the mountains, while unfriendly wet and cold weather prevails in the valleys and plains beneath the rather extensive fog or stratus cloud cover. If such fog layers were dissolved over larger areas, fog-free conditions could be maintained for many hours, as has been demonstrated in Soviet experiments. Thus, a large city or a metropolitan concentration could enjoy a few hours of sunshine.

'This is Utopia', I was told by a meteorologist with whom I discussed these perspectives: 'The effort would be much too large. Economy is not charity. No municipal administration would be prepared to throw money out of the window.'

There are no relevant cost–benefit analysis figures to counter such arguments. At best it could be pointed out that fog or stratus dissolution can produce some savings on heating and lighting and that, possibly, indirect savings could be made by reducing the number of working hours lost as a result of illness (colds, asthma, etc.) Nevertheless, it is likely that the balance would still show a deficit.

But is the 'value' of prevented asthma attacks really to be seen only in a reduction of working hour losses to the economy? Is the prevention of illness in children, housewives and old age pensioners of no value? Are values only measurable by doctor's fees, medicines and convalescence which would not arise if illness could be prevented? Even those who see matters only in terms of money must admit that illnesses do involve expenditure. But, to me, of much greater importance seems to be the principle that the health and wellbeing of people warrant expenses which do not bring an immediate return.

Municipal councils do not question the need for providing parks, green belts and other communal facilities such as baths, sport fields and recreational facilities which hardly pay for their costs. Why should one not consider cheering up citizens occasionally by a few extra hours of sunshine in autumn or winter through the dissolution of fog? In densely populated areas, the costs per person would probably not be much more than what is spent by people travelling at the weekend to a place where there is more sunshine.

Present technology is sufficiently advanced to realise such ideas at reasonable expenditure. It appears to me most important that the planning strategy of weather modification should place more weight on human aspects which cannot be expressed in monetary terms.

As things stand today, we have to assume that it will probably be a detergent manufacturer who, instead of placing a commercial on television, might advertise by providing a few hours sunshine for the citizens of the metropolis. Wouldn't a slogan like 'today's sunshine comes with the compliments of Sunbeam soaps' induce people to buy their product?

Ecological consequences

Usually, results of successful weather modification are within the natural range of weather fluctuations. Without statistical investigations it is therefore not possible to prove that such effects have taken place at all. We might assume that the relatively moderate changes in climatic conditions which can be produced by human intervention have no great influence on the animal and plant world. According to the few ecologists who have dealt with this problem, such a conclusion would, however, be wrong and harmful. Apart from extreme catastrophes the destruction of plant and animal societies is chiefly controlled by climatic conditions and not by single weather events. An increase of the *mean* annual precipitation by 10%, for example, would lead to gradual adaptation processes, even if the natural range of year-to-year changes is considerably larger. An ecosystem – a society of plants and animals within a given space, such as a forest – will not adapt suddenly and drastically to a new, slightly changed environment, but rather gradually, so that it would hardly be noticed by a nonbiologist. Only after decades would there be a marked change in the structure and assembly of species in the plant and animal society.[44]

Although we should assume on the basis of theoretical considerations that weather modification has ecological consequences it may be quite difficult to prove this in a particular case. Plant and animal communities continually undergo processes of adaptation to various natural variations and to changes which are inadvertently or intentionally caused by man. Weather modification is only one additional factor which combines with a number of others in producing environmental variations.[45] If we want to understand or even predict the specific consequences of weather modification in isolation from the other factors, intensive research must be carried out.

Unfortunately, such research had been neglected for a long time. Only after 20 years of commercial weather modification in the USA did relevant investigations commence in the mid 1960s and then only on a purely theoretical analytical basis. The work by Cooper and Jolly is so far the only comprehensive study of this problem. At the time of its publication (1969) not a single experiment had been carried out to study the possible ecological consequences of weather modification in detail.[46] Only during the 1970s have large-scale ecological investigations started. These are carried out within the framework of the Colorado Basin Project and the American Hail Research Project (NHRE), but no results are yet available.[47]

Cooper and Jolly searched the existing literature for indications of the effects of natural or experimentally induced environmental changes on plant and animal communities, so that they might infer from them the probable consequences of weather modification. They emphasised that this could only be done with reservations, since natural changes of precipitation are often coupled with temperature variations, while artificial augmentation of precipitation leaves the temperature unchanged. They stress the need for studying the effects of such changes on ecosystems in which there are many complicated interactions. Experiments which have been carried out with only one species cannot yield general ecological conclusions. In nature, many species compete with each other and improvements in the conditions for one particular species could in certain circumstances favour a competing species even more.[48]

If precipitation is augmented in an already moist region, as may be desirable in the interest of water supply to a hydroelectric station, there would be hardly any significant ecological consequences. Equally insignificant would be the consequences of relatively small increases of rainfall in very dry areas. The largest effects would be expected to occur in regions with moderate precipitation. There, a relatively small increase in precipitation might enable the establishment and the multiplication of species which under natural conditions could not thrive in that area. On moderately grazed natural pastures, where grazing animals influence selection of plant species, there would probably be larger changes than on ungrazed grassland.[49]

Temporary dissolution of fog is unlikely to have significant ecological consequences. Reduction in hail would further the distribution of those birds whose eggs and nests are often destroyed by hailstones. This would be of advantage to agriculture, since many birds feed on insect pests.[50]

Snowfall has been increased in mountains by orographic cloud seeding for the benefit of hydroelectric power stations. This might reduce the food supply for deer and elk, especially where high altitude dams and other human activities have already reduced the limited living space of wildlife. This could mean a serious reduction in animal populations and would probably be one of the first visible ecological consequences of tampering with the weather. However in managed hunting areas this problem could be solved by additional feeding in winter.[51]

It is improbable that weather modification at the present level would endanger any of the mammal species in North America; on the other hand, it is conceivable that some rare plants, which are now restricted to a small area with very specific environmental conditions, might lose their last habitat. Large-scale application of weather modification in the Near East could lead to the extinction of the most important wild strains of our grain plants. These contain valuable genetic properties which are needed for the breeding of new disease-resistant species. But, as Cooper and Jolly pointed out, American efforts to curb weather modification in the Arab countries at the same time as they are sponsored in the USA 'would rightfully be treated as a particularly intolerable form of neocolonialism'.[52]

The augmentation of precipitation for hydroelectric stations and dams increases river flow. The economist Crutchfield expects that this would be of advantage to commercial river fishing, particularly salmon, and also for sport fishing.[53] Cooper and Jolly are more conservative in their assessment. The effects might vary according to the fish species and local circumstances. Wherever large fishing interests are involved, the water should be examined in detail before large-scale weather modification projects are started. The release of large amounts of water into the river by hydroelectric power stations sometimes has serious consequences for aquatic life. Additional water from cloud seeding should therefore be used in a more flexible way with regard to the interests of the fisherman.[54]

It is unlikely that rain augmentation would produce catastrophic increases in plant pests, as their incidence is generally more sensitive to temperature than to precipitation. Increased rainfall during blossoming time could, however, infringe on cross-pollination by bees. Fungus diseases on plants might also increase, as humidity assists the spread of pathogenic fungi. Also weeds might increase considerably if soil moisture is raised. Some plant species which are not conspicuously active at present might spread and become a plague if much more rain were to fall.[55] Cooper and Jolly propose that in each weather modification project about 10% of the budget should be made available for ecological investigations. Before the start of a project stock should be taken of the environmental conditions and they should be monitored currently so that possible changes can be detected at an early stage.[56] The statement of a commission of the American Ecological Society which examined the ecological aspects of weather modification runs in a similar vein. Large-scale operations, which could influence climatic conditions over extensive regions should not be envisaged at the present state of knowledge, according to the commission, since the consequences cannot be foreseen.[57]

Among the substances used in weather modification, dry ice and liquid carbon-dioxide do not present ecological problems. Also, there seems to be no objection against the use of small amounts of liquid propane for the dissolution of fog. Whether silver iodide, the primary cloud seeding substance, is harmful to the environment is still not fully clear. Cooper reports on a series of studies made by different

129

groups of scientists. According to their results, no negative effects need be feared from the amounts envisaged to be used in the near future.[58] Ground crews working with smoke generators are exposed to high concentrations of silver iodide over prolonged periods. However, so far there have not been any reports of adverse effects.[59] Micro-organisms might respond more strongly to the effects of silver iodide than the multi-cell structures of larger forms of life. Investigations in connection with the Colorado project have shown that the seeding material concentrations found on the ground are still far below the threshold which could endanger the existence of bacteria in the ground.[60] Nevertheless, at the 1976 WMO Conference in Boulder, representative of the American National Science Foundation and others have spoken of 'continuing concern for longer detrimental effects of heavy metal on the environment'.[61]

Silver iodide is rather expensive. In the Soviet Union, where the annual use of 2.5 tons of silver for weather modification was felt to be a strain on the budget, the less expensive lead iodide (which has a similar crystal structure) is sometimes used in hail prevention operations.[62] Western countries have reservations about the use of this substance, which is considered to be toxic, particularly since too much lead is already injected into the environment by its use in petrol. As far as it is known, lead iodide has not yet been used for civilian purposes in the Western world, but it was used by the American military during operations in Vietnam.[63]

In the search for a cheap non-toxic and biodegradable seeding material, the western countries as well as the Soviet Union examined various organic compounds with appropriate molecular structures.[64] The tolerance of the environment to such substances, and particularly the question of whether they could increase the risk of cancer or cause deformities in newborn babies, will have to be investigated very carefully.

Weather makers in court

In the 1880s the high altitude regions of the American state of New York were hit by a prolonged drought. Duncan McLeod, a Presbyterian minister, one August Saturday organised a large procession with prayers. It was well attended. Only one farmer in the

130

area, Phinneas Dodd, did not join in the prayers. He was of the opinion that this was an uncalled for interference with the evolution of nature. Three hours after the service a heavy thunderstorm brought rain at last. Lightning struck Dodd's barn which burnt to the ground.

It is not difficult to imagine what happened next: the people applauded the minister and alleged that the lightning that struck Dodd's barn was God's punishment. Perhaps the minister should have countered such rumours or organised a collection for the unfortunate member of his community. But McLeod did nothing of the kind. Dodd was so enraged about the mockery that he finally took the case to court and demanded damages from the minister, arguing that the thunderstorm which had burnt down his barn was an immediate consequence of the prayer service organised by McLeod.

In court, the minister was in a difficult position. He had been celebrated for causing the rain and had repeatedly claimed in sermons to his flock that God had heard their prayers. How was he to state in court in the presence of many members of his community that a direct cause–effect relationship between the prayer service and the thunderstorm could not be proved? Soon everyone had the impression that McLeod was going to lose the case, when his defendant had a brilliant idea. After all, they had prayed for rain and not for thunderstorms. Thus his client was not responsible for the lightning which had destroyed Dodd's barn. On the basis of this argument, the case against the minister was dismissed.[65] Whether he would have been found guilty if floods instead of lightning had caused the damage remains an open question.

Weather modifiers who go to court nowadays to defend themselves against claims for damages are in a better position than Minister McLeod was in his time. They base their activities not on religious beliefs but on a scientific rationale. They can, with a clear conscience, assure their clients that they can modify weather within limits and at the same time defend themselves in court against a link between their activities and a specific weather event which caused damage to the plaintiff. This attitude proves to be successful, since, according to common law, it is not the accused who has to prove innocence but the plaintiff who has to prove guilt. So far, eight relevant court cases have passed through the courts in the USA and never has a plaintiff succeeded in convincing the judge of the causal connection between weather modification and damage.[66] Under given circumstances it is hardly likely that this will succeed in future. Even if there were a strong reason, based on scientific viewpoints, for assuming that damage had been caused by weather modification, this would generally not be sufficient legal proof.[67]

This unsatisfactory legal situation generates additional social hardship. In many cases the sufferers are small people who cannot afford to start a court case that promises little chance of success. But because they are small, they are much harder hit by damage than the clients of the weather modifiers: public utilities, agricultural corporations public authorities etc.[68]

Furthermore, ordinary courts and especially jury courts have little specific knowledge of the subject.[69] They must rely on statements by experts which, by American law, can be called in by both sides of the case. For this reason it has been proposed in the USA that an independent commission of experts in the relevant disciplines is created which can render specialist advice in court cases related to weather modification.[70] A study of the Stanford Research Institute recommended in connection with the large Colorado project that any reasonably founded claims for damage should be considered and settled by agreement outside the courts; also, that it should be a general rule that communities in the target area receive compensation for additional snow clearing operations and other inconveniences, without prolonged detailed negotiations.[71]

Court cases concerning weather modification bring up not only complicated problems of fact finding and proof but also basic questions of law that require clarification. As in many other areas, the development of legislation lags behind the technical progress. Most countries in which weather modification is carried out have few or no relevant laws. In the USA, any legal issues arising from it are generally settled at the discretion of individual states. A few of them have detailed rules and regulations concerning this matter, but these differ from state to state. Many have no such laws and one state (Maryland) has banned weather modification altogether. As long as there is no 'code of weather' anything not explicitly banned by law is allowed. The way in which general concepts of law are to be

132

applied to weather modification is subject to diverging opinions among judges, as has been shown by the few cases which have so far been treated.

Extremists of nature conservation reject any kind of weather modification as 'disturbance of nature's balance'. They often claim that silver iodide is toxic and should not be introduced into the environment. One legislator of Pennsylvania is of the opinion that silver iodide is harmful to animals and plants and restricts the reproductive faculty of eagles. Also dry ice (frozen carbon dioxide), which is occasionally used in weather modification, is claimed by him to be highly dangerous.[72] Pennsylvania has actually passed a law aimed at the prevention of damage to nature by weather modification.

The district authority of Ayr in Pennsylvania issued a local ruling against cloud seeding on the grounds of protecting the environment. A commercial weather modification firm which did not regard this ruling was taken to court and found guilty. A local authority has the right to impose a prohibition if it is of the opinion that there is any likelihood of damage to the environment. But when the 'Natural Weather Association' in another district of Pennsylvania tried to obtain an injunction against weather modification, they lost the case. A private plaintiff must *prove* that there is a risk of damage to the environment, and the Natural Weather Association was unable to do this.[73] The judge admitted that a land owner has the right to make decisions concerning the clouds and the water content above his land, but he maintained that in this case the plaintiff had not proved that weather modification had reduced the amount of precipitation which belonged, according to law, to him.[74]

Of greater importance than these rather amusing American disputes was a case heard in 1960 in New York State, when the City of New York wanted to increase its water supply by cloud seeding. The owner of a holiday resort in the target area of the weather modification operations brought a case against the City regarding disturbance of his business and demanded from the court to order the termination of artificial weather modification activities. The judge dismissed the complaint and stated in his summing up:

This court must balance the conflicting interests between a remote possibility of inconvenience to the plaintiff's resort and its guests with the problem of

133

maintaining and supplying the inhabitants of New York and surrounding areas with a population of about 10 million inhabitants with an adequate supply of pure and wholesome water. The relief which the plaintiffs ask is opposed to the general welfare and public good and the danger which the plantiffs apprehend are purely speculative. This court will not prevent a possible private injury at the expense of a positive public advantage.[75]

The principle of public interests taking precedence over private demands is also applied in many other situations, such as road construction or mining, but compensation is paid to people who have to give up their land. In the aforementioned case, compensation was not demanded and such a demand would probably not have been successful in view of the difficulties of proof. There is little doubt of the possibility of rain causing an infringement on the business interests of a tourist enterprise. This would justify an appropriate compensation, though, of course, not the banning of a project serving the public interest.

What serves the welfare of the public is, however, often a question much more difficult to answer than in the case against the City of New York. There was the Texas case in 1958, between a group of relatively poor ranchers and a commercial weather modifier who implemented a hail protection programme on behalf of a group of wealthy grain, fruit and vegetable growers. The ranchers claimed that rainfall on their land had been reduced by the cloud seeding. If meteorological experts had been called in, they would probably have said that at present it cannot be stated with confidence that hail prevention operations cause a reduction in the rainfall, but careful scientific observations rather show the opposite. However, the judge was satisfied with the statements made by the people of the region who said that cloud seeding had in fact reduced rainfall. He ruled that the aircraft of the weather modification company were no longer to operate above the region of the ranchers because 'the land owner is entitled to such rainfall as may come from the clouds over his property that Nature, in her caprice, may provide.'[76] A similar problem, whether a hail prevention project organised by barley growers had reduced rain for ranchers and vegetable growers, arose in the case of the San Luis Valley, which was described at the beginning of this book.

Obviously meteorologists are asked the impossible in the present

134

state of scientific knowledge if they are to decide whether a hail protection programme does reduce rainfall. And too much is expected from courts in the present state of legislation when they are asked to make decisions in such cases. But the problems arise now and cannot be removed by pointing out the inadequacies of dealing with them. They call for the creation of legislation which can be applied in practice.

Who owns the weather?

A scientist can bombard a nucleus with neutrons without asking the permission of the nucleus,

said Myron Tribus, Assistant Secretary for Science and Technology in the US Department of Commerce, during the second American Conference on Weather Modification, 1970.

He cannot engineer the environment without consulting the people who will be affected...Any procedure for planning weather modification activities must include provision for giving public and private interests a voice in the process. According to our historical political philosophy, we make important decisions by some adverse procedure whose main criterion for validity is that it gives all parties and interests an equal opportunity to be heard. We must not close the decision making process in weather modification. The results are too important to too many citizens and too many groups.[77]

In this general formulation the thesis of Tribus may well be accepted by any one. But, as soon as the general principles are examined in a little more detail, many questions arise which are difficult to answer. Who *owns* the weather and its 'elements', such as the clouds? Does an individual have a right to natural weather, uninfluenced by human intervention? Who determines weather policy? What guidelines are to be applied to these decisions? Who is concerned and must be heard? Who has the right to appeal against decisions and to which authority? Who can claim compensation for consequences of weather modification and how can the extent of such claims be assessed objectively? Who owns the 'products' of weather modification, such as additional rainfall which does not fall on the land of those who paid for the project?[78]

This book cannot provide the answers to these and similar questions which have not yet been answered by the legislators. It can

135

only point out some of the problems which will have to be dealt with in the future and some basic concepts which must be observed.

Weather is international: it does not stop at the borders of land owners or countries. The concept of a state or even an individual land owner 'owning' weather, and changing it at will is not realistic. Weather modification carried out above one area often has consequences which go far beyond the ground limits. Among the techniques which are now being applied, only the dissolution of fog is of local significance. The seeding of clouds for augmentation of rainfall, as we have seen, can have effects extending over several hundreds of kilometres. In a small country like Israel, this means that the effects extend across the national boundary. Modification of large-scale weather situations and climate, which may well be possible in the near future, affect even larger regions. Seen in this way, weather is the 'property' of the whole of mankind. The international legal problems which arise from this will be discussed in the final section of this book.

As for any nation's internal dilemma, the assumption might be made that weather is 'public property' which cannot be interfered with by individuals regardless of the interests of the community. The historical development, as we have described it, has led to the situation that in the USA weather modification is done predominantly on a private commercial basis, while scientific experiments are carried out by research institutions and public bodies. There is no central planning. 'Weather policy' evolves in a fragmentary and random fashion from unco-ordinated activities of various public and private sectors which have their own interests and problems at heart.[79] This is unsatisfactory. According to Fleagle and colleagues: 'Like virtually all prior technologies... weather modification has so far been left to develop without explicit consideration of society's needs and values... It is time that society claimed the driver's seat and got firm hold of the steering wheel before weather modification careers farther down the road. We have the opportunity if we but act promptly and wisely.'[80]

As matters stand in the USA today, weather modification is in many cases planned and carried out entirely in accordance with the commercial interests of the client (a public utility, a group of farmers, etc.) Other interests are only taken into account when they carry

136

sufficient voice and volume. The gearing of a 'weather policy' to the interests of the community would probably be best insured if the rules were made not by those who are interested in the 'products' of weather modification, nor by self-appointed authorities, but by a body responsible to the general public.

Any authority in a democratic country is in a way responsible to the general public as it operates under a government elected by the people. However, since weather affects every individual directly, a more direct participation in decision making by all concerned would be desirable. Naturally, it is not possible to call a plebiscite each time a particular aspect of weather modification arises, but this does not mean that all decisions need be left to a small number of experts who think they know best what is good for the interests of the general public. J. Eugene Haas writes that scientists, officials and weather modifiers are seldom correctly informed about the opinion of those living in the target area. However, the same officials often claim to know what the people think about it. Asked how they came to this knowledge, they refer to casual talks with local residents, from which they draw general conclusions. 'Repeatedly, we have found that even those atmospheric scientists who are very rigorous in the design and conduct of weather modification experiments find it convenient to ignore many of the most basic scientific procedures in reaching conclusions about social facts.'[81]

The more the public is taken into confidence, and the more people themselves participate in decision making, the smaller the difficulties and the unscientific bias against the weather modifiers. South Dakota elected 'commissioners' in each district to decide whether or not the district should take part in the rainfall augmentation and hail prevention projects which were carried out by state authorities. There was an advisory committee in each district which could call for an end to seeding operations if too much moisture was already in the soil, or if precipitation augmentation was considered undesirable for any other reason.[82] This attempt to secure participation of the population on a broad basis has met with great success.

Often, weather consultants and authorities carrying out weather modification experiments shroud their activities in a veil of secrecy.[83] In such cases, public reaction is frequently violent and, as in Quebec, weather modifiers are held responsible for undesirable natural weather events, even if, according to the view of meteorologists, they could not have arisen from the modification activities. In Pennsylvania, the bias against cloud seeding and the mistrust of authority was so strong that the army came under suspicion of having caused the drought in 1968 by secret seeding operations. The Department of Agriculture, which at the time carried out aerial photography to study the distribution of a plant disease in oak forests, had to terminate the survey, since the farmers, who thought that the low-flying aircraft were used for weather modification, started to shoot at them.[84]

In many instances, there is not just a bias but a real conflict of interests. 'Farmers and picnickers may often be at odds on the weather for Saturdays. Cities with snow removal problems and nearby ski resorts will also have conflicting desires.'[85] And of course there are the farmers' interests in weather differing according to the crops they grow.[86]

Experience gained in America shows that there is little chance that the pleas of potential opponents to a planned weather modification programme are heeded, unless they belong to a strong organisation which represents their interests.[87] An institutionalised participation of the people concerned could strengthen the legal position of potential sufferers against influential proponents of weather modification. However, this is not necessarily the best way to give maximum consideration to national interests which need not be identical with the interests of the inhabitants in the operational target area.

As we have seen, some weather modification operations may be generally approved by the people in the target area but nevertheless do not appear to be sensible when regarded from the viewpoint of the national economy. On the other hand there is the possibility that weather modification which brings benefits to a minority and damages a majority of people in the target area can still be helpful to the overall economy of the country. How is one to decide objectively in such cases which are the overall national interests? Who is to decide whether or not a project should be carried out?

The benefactors of rainfall augmentation are often outside the target area and may be, as in the case of New York, a large population. Cloud seeding in a river catchment not only benefits

138

hydroelectric power stations and their clients but, in a wider sense, the whole country, which can thereby reduce its import of oil as more power is created by water. At the same time, real damage may be inflicted on communities and individuals in the target area and in surrounding regions.

There must be ways of doing justice to people who suffer such damage without giving them the impossible task of proving their case. Finally, it has to be decided whether compensation is due only for material damage or perhaps also for other kinds of inconvenience, such as additional rain days suffered by the owner of a weekend house in the target area.

When weather modification is carried out on a commercial basis, as in the USA, questions arise as to who is authorised to practise it and whether a certificate of competence is required. Relevant regulations in the different states of the US vary on this point.[88] Also, there is no uniformity in the rules regarding the extent of the weather modifiers' third party liability and relevant insurance against it.[89] Nor is it clear how far a weather modifier is responsible to his client for compensation of damage arising from professional mishaps.

In most other countries weather modification is exclusively performed by scientists under contract with scientific institutions or authorities. This does not mean that in these cases the aforementioned problems are automatically solved. Scientists and officials do not possess secret recipes telling them exactly what is required for the common good, although they often pretend that they know. Besides, their activities are not always guided entirely by a desire of doing what is best for the community but, as with all other people, by personal interests and loyalties to their office or organisation.

Laws explicitly prohibiting commercial weather modification do not exist anywhere, as far as I know. In communist countries such a prohibition might be implied indirectly by the phrasing of the general economic laws and codes of conduct. In western Europe, it would be no breach of legal tradition if a new 'weather law', restricting the practice of weather modification exclusively to scientific and public institutions, were to be developed. Its application would probably cause procrastination and red tape, but this might be the lesser evil when compared with the consequences that could arise from a commercial handling of weather modification.

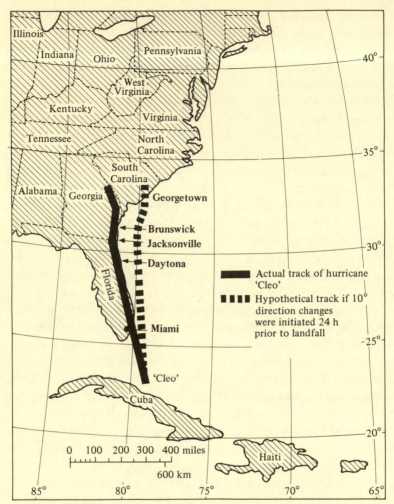

Fig. 20. Track of hurricane 'Cleo' and consequences if the track could have been altered by 10° in direction. (After W. R. D. Sewell (*b*).)

Legal and international problems of hurricane modification

In 1964, hurricane 'Cleo' moved along the east coast of Florida into Georgia and caused heavy damage to several large communities in both states. Professor Sewell writes: 'Had it been possible to alter the track of the hurricane by only ten degrees, the storm would have missed these communities. Unfortunately, it might have hit

140

Georgetown, South Carolina, instead. The total damages would have been considerably less than they were with the unmodified storm, but the incidence of losses would have been complete different.[90]

The possibility of actually modifying hurricane tracks in the near future creates a great number of national and international legal questions which, considering the size of this natural phenomenon, go far beyond the aspects discussed in previous chapters. The first question is who is to decide whether a hurricane should be deflected, especially when a change of track would increase individual suffering in the interest of an overall reduction of damage?

Sewell sees three possible choices for a decision-making body in the USA: The President; an office of the federal government; or a specially created authority.[91] The President might also base his decision on the advice of a special commission on which each of the states concerned could be represented. Decisions must be made promptly, without prolonged bargaining, otherwise the opportunity to alter the track of the storm may be lost altogether. In a study of weather modification in the public interest, Fleagle and his colleagues are of the opinion that this decision can only be made at the highest governmental level.[92]

Another equally difficult question concerns the criteria that should determine the decision whether or not to deflect a hurricane. To an outsider it might be obvious that a storm should hit a small town rather than a large city. However the inhabitants of the town in question are not likely to share this view. 'How many drownings in Key West balance x dollars of property damage in Miami'? asks the statistician G. W. Brier. 'How do you compare a politician's villa and a number of more modest dwellings of less influential people?'[93] Should a town for instance, be sacrificed to save the rocket launching site at Cape Kennedy or an important military base? Has a large industrial complex with high-cost investment preference over a residential area? Counted in dollars, the damage is much greater if the industry is destroyed, but if the storm levels the residential area, how is one to assess the human suffering – and how many electoral votes are going to be lost? Would a President agree to the 'rational' decision to deflect a hurricane if at the same time it meant the destruction of his own property and that of his family? Would his advisers make such a suggestion?

These kinds of problems are only rarely dealt with in the relevant literature, but nevertheless they do exist. The short time in which decisions have to be made on hurricane track deflection precludes any kind of participation by the potential sufferers. Opinions also differ as to whether such participation would be at all desirable and fruitful. It could be assumed that those who make these decisions are going to be responsible to the public after the event. This means that there must be basic guidelines and criteria on which such decisions have to be made, although no amount of legislative detail can cover all the possibilities that might arise in practice.

All writers on this subject agree that people who suffer from the deflection of hurricanes should be compensated. But who is to pay? Those who profited from the deflection? Or the state – which means the community of taxpayers? If a timely evacuation takes place, personal injury and death can often be avoided. But if they do occur, how are they to be compensated for? What consequences arise for insurance companies, and what kind of legislation is therefore required.

Professor Hunt of the University of Washington is of the opinion that the victim of a natural catastrophe such as a hurricane should always be compensated from public funds, no matter whether the hurricane followed a natural or a modified track.[94] But state help after catastrophes generally meets only part of the damage, while the victims of a deflected storm can rightly claim full compensation.

Other very complicated problems result from the fact that hurricanes bring important rainfall to many areas. In 1966, an extremely severe hurricane, 'Inez', occurred in the Gulf of Mexico and approached the coast of the USA. 'People had to be evacuated and the coast braced for damage' reports Walter O. Roberts, first director of the National Center of Atmospheric Research in Boulder, Colorado. 'The storm did not hit the coast in vulnerable areas however. At the time of its threat, few people would have balked at dissipating the hurricane, had we known how to do it. But the joke is that the hurricane produced, as it ran its course, widespread rainfall in the parched Altiplano of Mexico that assured crops and prosperity for a whole season and more.'[95]

Apart from possible international complications, the same problem also arises in southern and eastern regions of the United States. A lengthy, highly sophisticated report by the Stanford Research Institute, on problems of hurricane modification, seen from the viewpoint of decision theory, is notable for not considering this question at all, and thus presents completely distorted results.[96] Actually, in a hurricane the economic benefit of precipitation is often bigger than the damage caused[97] and a deflection of the storm means a deflection of the precipitation pattern. Will those who lose precipitation be entitled to compensation?

Fleagle and his colleagues bring up a number of further questions, which have not been answered. Who is responsible for the consequences when weather modifiers incur a professional mishap when treating a hurricane? What if the storm behaves in a way different from that expected when there is no obvious fault with the weather modifier? Can a government be held responsible if, when it decided against modifying a storm for apparently good reasons, it is subsequently found that the modification would in fact have brought benefits?[98]

The prospect that the USA might develop techniques of hurricane deflection and be able to mitigate storm damage has caused considerable interest in other affected countries, especially the Caribbean and South Eastern Asia. Obviously any experiments aimed at weather modification which affect the weather of other countries can lead to international complications and claims for compensation. It has been suggested that the USA should discuss these matters with the other nations involved and try to conclude some sort of agreement or treaty covering such questions.[99]

Countries which have bad relations with the United States may be suspicious of the hurricane modification experiments for several reasons. The USA have conducted experiments to influence hurricanes for many years. The Department of Defence was officially participating in their financing. American experts have repeatedly indicated the possibility of changing the track of hurricanes and have discussed the consequences that would arise therefrom. Gordon MacDonald has expressly pointed at the possibility of military application of such track changes. In view of the history of the America–Cuba relationship in recent years, it is not surprising that

Havana occasionally voiced the suspicion that the American military is trying to deflect hurricane tracks primarily in the direction of Cuba.[100]

Imagine a hurricane approaching the southern tip of Florida suddenly changing its track and devastating Havana. Although there is always the possibility of natural track changes in hurricanes which have not been seeded, surely the suspicion would arise that this event was the result of a malignant manipulation by the CIA. Such a suspicion could not easily be allayed, although M. Holden, professor of Political Science, points out that it would be difficult for a small developing country such as Cuba to verify the validity of such suspicions, as this would require a large number of experts in various relevant fields.[101] The mere fact that experiments do take place brings a new element of mistrust to international relationships and it is difficult, if not impossible, to eliminate it altogether by international agreements and control processes.

According to Professor Edith Brown-Weiss of Princeton University, the largest threat to peace is not posed by *actual* application of weather weapons, but by the suspicions and fears they cause in one country regarding their possible application by other countries. It will become possible to blame others for adverse weather and climatic changes. Once nations consider weather and climate as potential weapons, politicising of weather and climate can no longer be ruled out. This would add to the conflicts and tensions which arise from catastrophes, such as harvest failures as a result of drought or flooding.[102]

Truce on the weather front?

'Weather modification began as a military research and development program in Project Cirrus. Military services have continued research and development in the field since then and have undertaken some operational programs. The military research and development program has in many respects paralleled that in civilian community,' say Edith Brown-Weiss and W. Henry Lambright in a contribution to the 1974 Conference at Fort Lauderdale.[103] At first glance, the reported funding levels of military research and development in weather modification suggest modest involvement in the field. During the financial year 1974, total funding for weather modification research and development in the Department of Defense was $1 594 000.

As Fleagle and his colleagues point out, these figures constitute only a fraction of the Department of Defense's actual expenditure on weather modification.[104] They include neither the maintenance costs for aircraft put at the disposal of Project Stormfury, amounting to between $1 and 1.5 million per year;[105] nor the costs for rainmaking in the Vietnam War which, until the termination of operations in 1972, came to about $3.6 million a year.[106] Furthermore, the Department supports its Advanced Research Project Agency (ARPA) every year with about $3.3 million for reseach in global climate models which should make it possible to predict the effects of climate modification and to detect modifications carried out in other countries.[107]

Detailed information about this project, which runs under the code name 'Nile Blue', have not been released by the Department of Defense.[108] Finally, according to Brown-Weiss and Lambright, the expenditure for programme, installations and development techniques of weather modification in the US defence budget are widely scattered and the actual potentiality for development of new weapons is thus concealed. A combination of the presently separated elements might make it possible to develop new weapon systems.[109]

The Military has been the first, and for a long time the only American government authority dealing with weather modification and this has naturally shaped the development of this research area. As with other new techniques which might be usable as weapons, the defence authorities had a free hand in research and development without any centralised review of the desirability of having the new technology in the military arsenal.[110]

Civil authorities only recently became involved with the problems of weather modification after pressure had been applied to them by some Senators from drought-stricken states in the western USA. These politicians more or less forced the Weather Bureau and some other scientific bodies such as the National Science Foundation and the Academy of Sciences to take an interest in this originally 'non-scientific' field.[111] As a result of the Senators' actions, there are today a number of non-military authorities in the USA which deal with isolated aspects of weather modification and also apply them within the framework of their tasks. Yet, even today no 'nerve centre' exists which directs these activities and can develop a nationwide weather policy.[112] 'When there is no national policy on research and development in weather modification, there will be *de facto* policies that emerge from the particular power configurations within the bureaucracies at a given moment. The pulling and hauling of bureaucrats and technocrats may work out in the long run in the public interest of the country and the international community, but this would be more a matter of luck than of conscious, systematic planning.'[113]

Even within the military sphere there has been no uniform direction of research and development policy, according to the views of Brown-Weiss and Lambright. Fleagle and his colleagues report that Project Stormfury emerged only through the efforts of small groups in the Department of Defense and the Department of Commerce.[114] The aforementioned Pentagon specialist Canan sees this situation in a different way when he writes that, in the mid 1960s, all the military weather modification research was concentrated in the Cambridge Research Laboratory at Hanscom Air Base, Massachusetts. This laboratory is responsible to the joint Chiefs of Staff with participation of the CIA, an arrangement which was sanctioned by the White House under Johnson, and later under Nixon.[115]

As may be expected, there are legal experts, who maintain that military orientated research in this field is necessary and unavoidable. They argue that, even if one nation is of the opinion that an international ban of the application of weather modification warfare is desirable, it must consider the possibility of other countries preferring to develop this technique into an important and even decisive weapon.

Since there is no higher power which looks after the law and order of the international arena, each nation must rely on its own strength and that of its allies for the protection of its security and for its survival. As it cannot be certain that other countries would refrain from developing the weather weapon, each state has a good reason to develop this weapon.[116]

This is the well-known vicious circle logic of the cold war, according to which both sides invent more and more terrible weapons and develop them further, as they cannot know whether the other side etc., etc.... As a result of this, the great military strength of the two superpowers grows, and their national security – and with it the chances of survival of the whole mankind – is reduced year by year. According to J. S. Wiesner and H. F. York, who advised several American presidents on questions of international security:

Both sides in the arms race are thus confronted by the dilemma of steadily increasing military power and steadily decreasing national security. It is our considered professional judgement that this dilemma has no technical solution. If the great powers continue to look for solutions in the area of science and technology, the result will be to worsen the situation. The clearly predictable course of the arms race is a steady open spiral downward into oblivion.[117]

Of course, it must be assumed that the Soviet Union also proceeds with the development of weather and environmental weapons on the basis of the same considerations. But there, such activities are shrouded in a dense veil of secrecy while in the USA, by American tradition, such questions are openly discussed. There may also be other countries which study the military application of weather modification, but nothing is known about that.

In the USA at a relatively early stage, other voices have also been heard, particularly in scientific circles. On the one hand, some people are worried that a strong participation of the Military in weather modification would distort the international co-operation of meteorologists so important for the daily routine of weather services and for further research and, on the other, it could be that the military application of relatively harmless techniques, such as the dissolution of fog, or the increase in precipitation to disturb enemy supply routes, are only the first steps on a fateful path to more and more terrible weather and environmental weapons. A similar situation existed during the First World War, where the military application of tear gas was the first step towards the use of much more poisonous gases.[118]

In the 1960s, a commission in the American Academy of Sciences under the chairmanship of Gordon MacDonald studied the prospects of weather modification. It called on the President and Congress to make an official declaration to the world that the USA would use weather modification only for peaceful purposes and would support international co-operation in this field. However, the State Department advised at that time (1966), that it was 'too early' for such a step.[119]

MacDonald was not satisfied. Obviously, he was deeply troubled by the thought that the Military might implement ideas which he himself may have given to them when he was Vice-President of the Institute for Defense Analysis. In his afore-mentioned contribution to a book on the dangers of modern weaponry, published in 1968,[120] he tried to alarm the public about the terrible possibilities of weather and environmental warfare. He was also the initiator of a communiqué of American scientists, signed, among others by 22 Nobel Prize-winners, against the application of the weather weapon in Vietnam.[121]

Probably as a result of MacDonald's efforts, the American Senate became interested in these questions. The principal spokesman in these matters was Senator Claiborne Pell, a Democrat from Rhode Island, who, on several occasions, demanded information on the application of cloud-seeding in Vietnam, from representatives of the Pentagon, but for a long time received only evasive answers.

In March 1972, Pell tabled a motion in the Senate asking that the USA try to conclude an international agreement on the prohibition of environmental warfare. This move was supported among others by prominent politicans such as ex-Vice-President Humphrey, the Democratic Presidential candidate for 1972, McGovern, and Senator Edward Kennedy. After discussions in the foreign relations committee of the Senate, it was adopted by 82 votes to 10. The proposed treaty envisages the prohibition of all forms of weather and environmental warfare, including cloud seeding, hurricane track modification, climate modification, and the artificial triggering of earthquakes or flood waves (tsunamis), as well as all research and development activities for military purposes in these fields.[122] The proposal contains, in fact, the whole list of weather weapons and environmental weapons compiled by MacDonald. There are no guidelines about controls of observation of this treaty in the draft. According to Edith Brown-Weiss, it would be possible to detect at least the larger infringements by way of observations from weather satellites and ground stations, and by comparison with outputs from weather forecast computers.[123]

Similar discussions in the USSR have not been carried out publicly

148

and we have to rely on second-hand information. The well-informed Director of the Stockholm International Peace Research Institute (SIPRI), Frank Barnaby, writes that 'A leading Soviet proponent of the prohibition of geophysical weapons is academician Federov, chief adviser to the Soviet Government on problems of environmental warfare.'[124] Federov, in his contribution to the book edited by Hess, expressedly supports international co-operation in the field of weather and climate modification. 'There is hardly any need to argue that under these conditions all mankind must present a united front to the world around it. There is no other alternative. There is nothing absurd in dreams of a "struggle for meteorological mastery". Soviet scientists harbour a deep concern about cessation of the arms race, disarmament, and the maintenance of peace...'[125]

Perhaps it can be taken as a result of the parallel efforts of MacDonald, Fedorov and others of the same opinion, that, during the Nixon–Brezhnev meeting at Vladivostok in July 1974, among other things it was agreed to start a Soviet–American exchange of thoughts on these questions. The Soviets did not wait for the result of his exchange. Already in September 1974, they unilaterally tabled at the UN General Assembly a draft convention prohibiting modification of the environment, including weather and climate, 'for military and other purposes incompatible with the maintenance of international security, human wellbeing and health'.[126]

'To the Pentagon, this was nothing more than a typical Russian ploy', writes Canan. 'Whether the Soviets were ahead or behind in research and development of ways to prostitute the elements, they stood to gain. If they were ahead, enactment of prohibitive international law would permit them to stay ahead. If they were behind – as was probably the case – they could simply break the law and work to catch up. The latter hypothesis presupposed, of course, that the Pentagon would not break the law, and would disband its R & D.'[127]

In the prevailing international climate of 1974/75, however, the 'hawks' on both sides could not get the upper hand. The White House as well as the Kremlin were interested in a relaxation of international tensions, not least because of pertaining internal political circumstances.

In any case, both superpowers unanimously tabled proposals in

149

1975 at the Disarmament Conference in Geneva concerning a treaty for the prohibition of environmental warfare. In it, it is envisaged 'not to engage in military or any other hostile use of environmental modification techniques having widespread, long lasting or severe effects as the means of destruction, damage or injury to another state party'. Environmental modification techniques are defined as 'the deliberate manipulation of natural processes' including weather, climate, earthquakes, tidal waves, ocean currents, and any upset in the ecological balance of a region. This proposal is fairly moderate compared with earlier drafts, and therefore perhaps more realistic. It was signed in May 1977 in the presence of the UN General Secretary, Kurt Waldheim, by the foreign ministers Cyrus Vance and Andrei Gromyko and representatives of other nations which participated in the Geneva Arms Limitation Conference. The treaty will become effective once it is ratified by at least twenty nations.

The significance of this treaty's contribution to international efforts of reducing the arms race should not be overestimated. Certainly, adherence to it would imply that military powers refrain from using presently existing techniques of relatively minor significance, such as cloud seeding for the purpose of disturbing enemy supplies; and more importantly it would also mean their foregoing the use of more atrocious terror weapons that are probably not yet developed or, even if they already exist, have not yet been tested as their reliability may be in doubt and one cannot be sure whether their use would not affect their own country. However, the worst threat to today's society are not the hypothetical environmental weapons of the future but the already existing atomic weapons, the delivery systems of which are more and more perfected and automatised. As long as there is no real limitation of the nuclear arms race, treaties on fringe issues, such as environmental warfare, could possibly help to distract world attention from the fact that no progress is made on the really burning questions.

At the same time it would be equally unjustified to deny this treaty any concrete value. As it concerns terrible weapons that cannot be tested and are probably unreliable, there is hope that it will be adhered to, just as the prohibition of the use of poison gas was generally adhered to during the Second World War. Moreover such

150

a treaty is an essential precondition for fruitful worldwide co-operation in the peaceful uses of weather and climate modification.

World weather authority – a step towards world government?

It is clear that a long range program of weather control and climate modification can have a direct bearing on the relations between nations. It can aid the economic and social advancement of the less developed countries, many of which face problem associated with hostile climates and serious imbalances in soil and water resources and, quite importantly, *it can serve to develop common interests among all nations and thus be a stimulus for new patterns of international co-operation*

reads the report of the MacDonald Commission of the US Academy of Sciences.[128]

In the same vein Professor Federov writes: 'The problem of global and regional climate modification is essentially international and requires the common efforts and co-odinated planning of all nations.'[129]

At the UN Environmental Conference in Stockholm in 1973 it was agreed in principle that the atmosphere belongs to all people in the world and must not be modified unilaterally by one country without regard to the interest of other nations. At present, however, cloud seeding techniques can have remote effects which reach far beyond the borders of smaller countries. Experiments aimed at the modification of large-scale weather situations and climate, and those intended to change the tracks of tropical storms, are undoubtedly going to affect many countries. It is obviously necessary to create an internationally accepted regulating mechanism for such activities. In 1967, W. O. Roberts of the National Center for Atmospheric Research in Boulder, Colorado, pointed out that such questions should be dealt with *as soon as possible*, as it is easier to reach an international agreement as long as the scientific-technological development is still flexible.[130]

Such efforts would probably not have been realistic before the treaty on the prohibition of weather and environmental warfare had been signed,[131] but now that the necessary preconditions exist, it should no longer be delayed.

'I strongly urge that an international body be formed to consider possible areas of conflict, and to suggest the structure of future international authorities or agreements to deal with climate-related conflict situations before they arise and before their potentially irreversible consequences are felt,' writes Stephen H. Schneider, climatologist at the National Center for Atmospheric Research in Boulder: 'The recent stirrings in the United Nations to ban weather warfare are an encouraging beginning, but a much more comprehensive mechanism is needed. A major threat to international peace and security could conceivably come even before the end of this century, not only from the much feared spectre of nuclear war, but also from the less known, slower acting, and possibly more ominous potential of an altered climate, whether modified by accident or design, for peaceful purposes, or for war.'[132]

The over-riding guideline in any international agreement should be that experiments, as well as all operational projects, that might have a bearing on the weather or climate of other countries, should only be carried out with express agreement of these countries. In any case, wherever possible, such projects should have the participation of all countries concerned.[133] It will probably be necessary to create within the framework of the World Meteorological Organization (WMO) an international institution whose highly qualified experts examine any planned projects and discuss any objections against them.[134] Since small and economically weak countries usually do not have sufficient expertise to study these projects and their possible consequences in detail, it is all the more important that such an international body recruits competent experts who do not represent the view of their home countries, but rather the interests of mankind and especially those of less influential countries. Apart from meteorologists, such a body of experts must include economists, ecologists, doctors, lawyers and others who are in a position to assess the effects of weather and climate modification projects on the various aspects of natural life and society and – if possible – predict them.

Projects whose consequences extend beyond local boundaries should be carried out only after they have been assessed and approved by such a world weather authority. Working papers and results of operations should be available to all member states. The creation of special supervisory bodies should be considered which – similar to the inspectors of the International Atomic Energy Agency in the field of reactor technique – supervise the adherence to a 'world weather treaty'. Since weather modifications are far less fraught with 'industrial secrets' and commercial interests than the nuclear technology, the realisation of international control on weather modification should not present great problems.

The creation of such a supervisory mechnism could contribute considerably to defusing the 'politicising' of weather, as feared by Edith Brown-Weiss. Under the eyes of expert inspectors it would hardly be possible to camouflage subversive projects to make them look like operations for improving the local climate. On the other hand, accusations that a weather catastrophe was caused by the manipulation of a hostile country could be objectively examined and unjustifiable suspicions eliminated. A continuation of research and development work which is primarily serving military purposes might be considerably more difficult under such conditions.

Finally, agreements and rules should be made regarding the payment of compensation, within the framework of an international treaty to countries suffering disadvantages in cases where the outcome of a project was different from that expected. It may sometimes be difficult to find out what are freaks of nature and what are the consequences of human intervention. Nevertheless, a way of making a claim should be found if in all probability damage has been caused. Perhaps a fund could be created, within the framework of a world weather authority, to which all member states contribute, and which can be used to meet such claims.

The significance of an international ruling on weather modification must not be overestimated. In today's world, international organisations are rather weak, certainly weaker than the individual states participating in them. Within a world weather authority, the tune would most likely be called by the nations capable of developing new techniques and naturally wanting to use them for their own benefit. Those that might suffer from such projects are less likely to have much influence. Nevertheless, even an imperfect solution is better than a free-for-all in a subject which is not covered by international law.

Treaties among individual states, however detailed, cannot fully meet the problems of large-scale weather and climate modification. Basically, this is an issue on which the whole of mankind must decide. Under the present world political system, individual countries take

153

decisions of greatest significance for the evolution of the environment and even for the survival of mankind. These decisions are primarily governed by national interests, while considerations of the common benefit of all nations take only second place. A world system composed of national units is not the best solution for the application of either large-scale weather modification or many other techniques which require worldwide planning.[135]

Thus, the considerations regarding future prospects of weather modification lead to one of the biggest problems of our time. Scientific technological development proceeds at a breath-taking speed and creates more and more opportunities which can only be taken advantage of on a worldwide basis. More and more problems arise that must be solved on a similar scale. Especially the arms race had brought on the need for a new world order in which wars are no longer possible. However, neither existing institutions nor the ideas and goals of individuals have kept step with this development. Even in relatively small countries, like Northern Ireland, Cyprus or the Lebanon, the problem of human co-existence between majority and minority cannot be solved in a way that satisfies both parties. Under such conditions it is not surprising if existing nations refrain from handing their power to a 'world society' in which a sovereign people could easily become a suppressed minority.[136]

On a higher level, mankind faces problems similar to those at the end of the middle ages, when the then feudal state order became too restrictive. It was a long, painful and bloody process which finally resulted in the system of centralised national states, which could deal better with the new circumstances. With it came the emergence of national consciousness, which was hardly in being in Europe during the middle ages.

'World peace is not the "Golden Age" but its coming is manifest in a gradual transition of the present existing foreign policies into world internal policy' explained the physicist and philosopher, Carl Friedrich von Weizsacker in 1964, when he received the Peace Prize of the German Book Trade

Not even 100 years have gone since German states fought each other for the last time. This is hardly imaginable by the present generation. We must hope that, to those who will be young in 100 years, the past quarrels between Germany and France, and even the possibility of a war between America and Russia, should be just as unimaginable as the political condition in Germany which was terminated by the wars of 1866 and 1870.[137]

We live in a transitional period in which a consciousness of global citizenship must develop. If the transition is equally lengthy, painful and bloody as that from the feudalism to the present, then there is the danger that mankind will annihilate itself in the process – either by a worldwide nuclear war, or more generally by progressive destruction of its environment.

If mankind wants to survive, the power of individual states will have to be curbed progressively and the share of influence of international institutions will have to be enlarged. Scientific-technological development brings about facts which do not allow an alternative choice. If there is to be a future at all, it will not belong to the strongest individual states but rather to the hitherto relatively weak worldwide organisations. Although these are now still dominated by member states, they may become the nuclei of a new kind of worldwide planning and co-operation.

The idea that a 'world government' will evolve not so much on the political plane but rather 'through the back door' on the level of specialised international organisations must sound Utopian. Anybody familiar with the present mode of operations in the World Health Organisation, the Food and Agriculture Organization, UNESCO, the International Atomic Energy Agency and other similar institutions will have a difficulty imagining that these organisations, which are at present used politically by member states, should gradually grow into 'world ministries' and take over certain sovereign rights from member countries. But, unless such Utopian-sounding solutions are pursued, there appears to be no way at all out of the crisis in which the present world order is submerging and which can no longer deal adequately with the technological scientific development. The 'realism' of armament all over the world, and the narrow-minded forcing of national interests can only lead to a world catastrophe.

In this Utopian perspective – the only remaining hope for a survival of mankind – a 'world weather authority' could play an important and exemplary role. There are good chances for just such an institution becoming a 'world authority of a new type' and of the realisation of new ways of international co-operation, of the kind

155

outlined in the MacDonald Report of the American Academy of Sciences.

The daily routine work in meteorology requires co-operation of the weather services in all states, large and small. Only a weather map without blank areas is helpful in the preparation of a reliable weather forecast. For this reason, meteorologists are usually endowed with a certain amount of open-minded readiness for international co-operation which would be a great asset in favour of the smooth functioning of a world weather authority.

Weather modification and the manufacture of equipment required for it is not big business. In contrast to other branches, such as nuclear technology, there are no mammoth-sized multinational corporations that would derive an immense profit from the military and peaceful uses of weather modification, and would exert tremendous influence to sabotage international agreements which run counter to their commercial interests. Weather and environmental weapons play a much lesser role than nuclear weapons and there is still hope for a curbing of the military orientated thinking in this research area.

Seen from this angle, there appear to be better chances for a successful functioning of a world weather authority than for that of the International Atomic Energy Agency (IAEA), whose scope is restricted right from the beginning in two ways: first, by nuclear powers having a privileged position within this framework as they are not under the control of IAEA and, secondly, because a number of states such as India, South Africa, Brazil, Israel and others, by not signing the non-proliferation treaty left their options open to become nuclear powers and have no control agreement with the Agency.

In a world weather authority created after the ratification of a treaty on the prohibition of weather and environmental warfare, there must be no difference between 'weather powers' and other countries. The authority would have to exert control on all countries in the same way, guide the peaceful application of weather modification proceeds according to global planning and make sure that it is not abused for military purposes. If there are no special privileges for 'weather powers' then there is less temptation for other countries to reserve the right of development of weather weapons. Thus, there is a greater chance that, by and by, all countries of the world would submit to a world weather authority.

At present all this looks like wishful thinking. But the possibility to move along this way is open, and, in spite of well-founded scepticism, we may hope that the idea can be realised.

The planning of experiments for modification of weather and climate on a large scale will naturally continue. These experiments cannot be arranged in a way that certain regions, such as the Soviet Block countries or the participants to NATO, remain unaffected. Results of these experiments can only be assessed correctly if data from all affected countries are analysed. Whether such experiments are useful and sensible is still subject to debate and must be clarified by international preparatory work. But, if the results of these studies should be positive, there is all the more reason that future operations are handled by international co-operation from the beginning. Such common global efforts would go a long way towards easing mistrust and tension between nations, and towards the creation of interests common to all countries. In this manner, weather modifiers might even exert some positive influence on the world's political climate and contribute a little to finding a way out of this hopeless world of the 'realists' and towards the 'Utopian' situation in which there are no more wars.

Notes

Weather modification: prospects and problems

1 (a) J. Eugene Haas in W. N. Hess (editor), *Weather and Climate Modification*, New York, 1974, p. 791.
2 *New Scientist*, 22 March 1973, p. 674.
3 Robert D. Elliot in (a), p. 48; Horace A. Byers in (a), p. 22.
4 (b) W. R. D. Sewell in W. R. D. Sewell (editor), *Modifying the Weather*, University of Victoria, 1973, p. 1.
5 Robert G. Fleagle, James A. Crutchfield *et al.* in *Weather Modification in the Public Interest*, University of Washington, 1974, p. 28.
6 M. T. Charak and M. T. DiGiulian, *Weather Modification Activity Reports*, NOAA, Rockville, 1974, p. 3.
7 E. K. Fedorov in (a), p. 394.
8 E. J. Smith in (a), p. 422; A. Gagin and J. Neumann in (a), p. 454.
(c) L. Facy in *Proc. WMO/IAMAP Scient. Conf. on Weather Modification*, Tashkent, 1973, WMO, 1974, p. 8.
9 A Working Group on Cloud Physics and Weather Modification of the World Meteorological Organization (WMO) studied the present state of knowledge and possibilities of weather modification and submitted a report to the WMO Executive Committee which was accepted. See:
(d) *Proc. Second WMO Scient. Conf. on Weather Modification*, Boulder, Colorado, 1976, WMO-433, 1976, p. xv.
10 J. E. Haas in (a), p. 794.
11 (b), p. 41.
12 (e) A comprehensive treatment of the possibilities of weather weapons and environmental warfare is given by G. J. F. MacDonald in N. Calder (editor), *Unless Peace Comes*, London, 1968, pp. 165–183.

1 Scientific and technical background

1 G. Fischer, *Umschau*, **75** (1975), p. 266.
2 D. Rousseau in R. Glowinski and J. L. Lions (editors), *Computing Methods in Applied Sciences*, Second Int. Symp., Dec. 1975, Springer, Heidelberg, New York, 1976, pp. 331–346.

3   L. Kletter, *Naturwissenschaftliche Rundschau*, **15** (1962), p. 90.

4   Joanne Simpson and A. A. Dennis in (*a*), p. 241.

5   Joanne Simpson in (*a*), p. 238.

6   A. Gagin and J. Neumann in (*a*), p. 465.

7   P. Grosberg, *The Changing Role of the Textile Engineer*, Leeds University Press, 1965, p. 11.

8   Quoted by H. R. Byers in (*a*), p. 15.

9   Joanne Simpson and A. A. Dennis in (*a*), p. 232; G. MacDonald in (*e*), p. 193.

10   A. Gagin and J. Neumann in (*a*), p. 457.

11   *Ibid.* p. 195.

12   B. J. Mason, quoted in (*a*), p. 227; in a review of (*a*) published in *Weather* (1975, p. 202) Mason, while maintaining his generally very negative view of weather modification, gives credit to the work done in Israel. Gagin's and Neumann's 'sensible and critical approach', he says, 'lends credibility to their claims for producing increases of precipitation of about 15 per cent'.

13   G. W. Brier in (*a*), p. 206.

14 (*f*) D. Rottner *et al.* in *Jour. Appl. Met.*, **14** (1975), p. 939; this special issue of the journal summarises the results of the Fort Lauderdale Conference on Weather Modification.

15   Joanne Simpson and A. A. Dennis in (*a*), p. 269.

16   S. A. Changnon Jr. in (*c*), p. 415.

17   WMO statement quoted in (*d*), p. xv.

    (*g*) See also the pioneering contribution of M. Tribus, *Science*, **168** (1970), p. 201.

18   D. Atlas in (*d*), p. 215.

19   J. Warner in (*c*), p. 43.

20   Joanne Simpson in (*f*), p. 663.

21   A. Berson, *Wissenschaftliche Luftfahrten*, II, Braunschweig, 1900, p. 192; A. Wagner, *Sitzungsber. der oesterr. Akad. der Wiss., Naturwiss.-mathem. Klasse*, Vienna, 1908.

22   A. Wegener, *Thermodynamik der Atmosphaere*, Leipzig, 1911.

23   T. Bergeron, *Geof. Publ.*, **5**, No. 6 (1928); T. Bergeron, *Assoc. Met., IUGG, 5th Gen. Assembly*, Lisbon, 1933, p. 156; W. Findeisen, *Met. Zeitung*, **55** (1938), p. 121; *ibid.*, **56** (1939), p. 365.

24   T. Bergeron, *Berichte des Deutschen Wetterdienstes*, **12** (1950), p. 225.

25   H. G. Houghton, *Bull. Am. Met. Soc.*, **19** (1938), p. 152; G. C. Simpson, *Quart. Journ. Roy. Met. Soc.*, **67** (1941), p. 99.

2 Methods and applications

1   H. R. Byers in (*a*), p. 21; his contribution on the history of weather modification provided much of the information used for writing this chapter; see also R. D. Elliot in (*a*), p. 45, on the history of commercial weather modification; I. Langmuir, *Collected Works*, vols. **10**, **11**, New York, 1961; *Trans. N. York Acad. Sci.*, **14**/I, (1951), p. 40; V. J. Schaefer, *Science*, **104** (1946), p. 457; *Bull. Am. Met. Soc.*, **49**, p. 338; B. Vonnegut, *Chem. Revs.*, **44** (1949), p. 177.

2   M. Holden Jr in (*b*), p. 277.

3   Tor Bergeron, *The Problem of Artificial Control of Rainfall on the Globe*, *Tellus*, **1** (1949), No. 1, p. 32; No. 3, p. 15.

4 (*h*) For a summary see J. D. Lackner, *Precipitation Modification*, US Nat. Tech. Info. Service, PB 201534, US Dept of Comm., 1971, Sect. V.

5   B. N. Leskov in (*c*), p. 143.

6   B. A. Silverman and A. I. Weinstein in (*a*), p. 378.

7 (*i*) H. A. Chary *et al.*, *7th Ann. Rep. on AWS Modif. Prog.*, AWS-TR-74-254, July 1974, p. 12.

8   B. A. Silverman and A. I. Weinstein in (*a*), p. 384.

9   *Ibid.*, p. 371.

10   L. Facy in (*c*), p. 8.

11   E. Sauvalle in (*d*), p. 397.

12   B. A. Silverman and A. I. Weinstein in (*c*), p. 21; A. I. Weinstein *et al.*, A modern thermal fog dispersal system for airports, *Int. Symp. for Artif. Weather Modif.*, IUGG, Grenoble, 1975.

13   L. Facy in (*c*), p. 369.

14   R. A. Sax in (*f*), p. 667.

15   H. A. Chary in (*i*), p. 10.

16 (*j*) I. V. Litvinov in J. Sedunov (editor), *Cloud Physics and Weather Modification*, English Translation, Jerusalem, 1974, p. 24; A. I. Weinstein and B. A. Silverman in (*a*), p. 367.

17 (*k*) G. Mueller *et al.*, *Preprints of the Fort Lauderdale Conference on Weather Modification*, 1974, p. 286.

18   B. Federer, *Umschau*, **67** (1967), p. 707.

19   N. S. Bibilaschwili *et al.* in (*c*), p. 333.

20   F. P. Parungo *et al.* in (*d*), p. 505.

21   J. Neyman and E. Scott, *Journ. Am. Statist. Ass.*, **56** (1961), p. 580.

22   J. Warner in (*c*), p. 43.

23   P. Squires and E. J. Smith, *Austr. Journ. Scient. Res.*, Ser. A., **2** (1949), p. 232; E. J. Smith, *Journ. Appl. Met.*, **9** (1970), p. 800.

24   J. Warner in (*c*), p. 43.

25   A. Gagin and J. Neumann in (*d*), p. 199.

26   J. Neyman and E. Scott, *Science*, **163** (1969), p. 1445.
27   S. A. Changnon Jr in (*c*), p. 415.
28   J. Flueck, *Final Rep. on Project Whitetop*, part V, University of Chicago, 1971.
29   J. D. Marwitz in (*d*), p. 85; R. Biondini in (*d*), p. 159.
30   L. O. Grant in (*f*), p. 659.
31   R. D. Elliot in (*a*), p. 55.
32   *Ibid.*, p. 57.
33   *Ibid.*, p. 81.
34   *Final Rep. Adv. Comm. on Weather Control*, US Government Printing Office, 1958.
35   L. O. Grant in (*f*), p. 660.
36   M. Tribus in (*g*), p. 206; L. O. Grant in (*a*), p. 282.
37   J. D. Marwitz in (*d*), p. 861; L. Vardiman *et al.* in (*d*), p. 41.
38   *Res. Act. CSIRO, Division of Cloud Physics*, June 1976, p. 28; E. J. Smith in (*a*), p. 442.
39   L. O. Grant in (*a*), p. 308.
40   On the San Juan experiment see also J. D. Marwitz in (*d*), p. 85; W. A. Cooper and C. P. R. Sanders in (*d*), p. 93.
41   L. O. Grant in (*d*), p. 308; L. O. Grant in (*f*), p. 660.
42   R. D. Elliot and W. Schaffer in (*k*), p. 496.
43   P. V. Hobbs in (*f*), p. 738.
44   H. K. Weickmann in (*a*), p. 318; M. Tribus in (*g*), p. 205.
45   Joanne Simpson and A. A. Dennis in (*a*), p. 229.
46   Joanne Simpson in (*f*), p. 664.
47   J. F. Hitschfield in (*f*), p. 657.
48   J. D. Lackner surveys this matter in (*h*), Sect. IV.
49   E. E. Kornienko *et al.* in (*c*), p. 51.
50 (*l*) L. J. Battan, *Bull. Am. Met. Soc.*, **58** (1977), p. 16.
51   M. V. Buikov, E. E. Kornienko and B. N. Leskov in (*d*), p. 141.
52   L. J. Battan in (*l*), pp. 5 and 19.
53   E. K. Fedorov in (*a*), p. 395.
54   Joanne Simpson in (*a*), p. 247.
55   *Ibid.*, pp. 253, 255 and 259.
56   Joanne Simpson and W. L. Woodley in (*f*), p. 738; Joanne Simpson and A. A. Dennis in (*a*), p. 279.
57   Joanne Simpson in (*f*), p. 665.
58   Joanne Simpson and A. A. Dennis in (*a*), p. 240.
59   W. L. Woodley, Joanne Simpson *et al.* in *Preprints, Conf. on Planned and Inadvertent Weather Modification*, Champaign-Urbana, October 1977, p. 206; W. L. Woodley, Joanne Simpson *et al.*, *Science*, **195** (1977), p. 735.
60   *Ibid.*, p. 735; W. L. Woodley *et al.* in (*d*), p. 151; R. Biondini in (*d*), p. 27.
61   A. M. Kahan *et al.* in (*d*), p. 27.

62 (m) *Summary Report, Weather Modification, Fisc. Year 1972.*
NOAA, Washington, 1973, p. 24; K. R. Biswas *et al.*, *Journ.*
Appl. Met., **6** (1967), p. 914; Joanne Simpson and A. A. Dennis
in (*a*), p. 243; and other contributions in (*d*), pp. 3–24.

63 H. Byers in (*a*), p. 27.

64 Joanne Simpson and A. A. Dennis in (*a*), p. 270.

65 H. Byers in (*a*), p. 6.

66 I. W. Litvinov in (*j*), p. 30; N. I. Wulfson and M. L. Lewin in
(*c*), p. 255; L. J. Battan in (*l*), p. 8.

67 Joanne Simpson and A. A. Dennis in (*a*), p. 272.

68 W. M. Frank and W. M. Gray in (*k*), p. 193.

69 C. S. Downie and R. A. Dirks in (*d*), p. 561.

70 Joanne Simpson and A. A. Dennis in (*a*), p. 272; J. Huff in (*f*),
p. 978.

71 (n) J. Sedunov in *WMO Comm. for Atm. Sciences*, CAS-VI/Inf. 7,
11 January 1974, p. 10

72 M. V. Buikov *et al.* in (*d*), p. 135.

73 E. S. Artsybaschew *et al.* in (*c*), p. 265.

74 E. G. Bowen, *Science Journal*, Aug. 1967; R. D. Elliot in (*a*),
p. 48.

75 A. Gagin and J. Neumann in (*a*), p. 462.

76 J. D. Lackner in (*h*), Sect. VI.

77 *Ibid.*, p. 19.

78 H. Byers in (*a*), p. 14.

79 (*m*), p. 24.

80 (o) R. G. Fleagle *et al.* in *Weather Modification in the Public
Interest*, University of Washington, 1974, p. 65.

81 G. J. F. MacDonald in (*e*).

82 *Weather and Climate Modification – Problems and Prospects*,
Nat. Acad. Sci./Nat. Res. Council, 1350, Washington, 1966. In
this commission chaired by Gordon J. F. MacDonald, the cloud
physicist James E. MacDonald played an important role.

83 G. J. F. MacDonald in (*e*), p. 170.

84 *The Pentagon Papers*, Boston, 1971, p. 281.

85 *New Scientist*, 4 May 1972, p. 281.

86 *New Scientist*, 12 October 1972, p. 110.

87 A comprehensive report on this hearing is given in *Science*, **184**
(1974), p. 1059.

88 (p) J. W. Canan, *The Superwarriors; The Phantastic World of
Pentagon Superweapons*, New York, 1975, p. 358.

89 (*m*), p. 39.

90 J. A. Crutchfield in (*b*), p. 205.

91 (q) R. G. Fleagle in Fleagle (editor), *Weather Modification –
Science and Public Policy*, University of Washington, 1968, p. 9.

92 H. Byers in (*a*), p. 32.

93   *Ibid.*; H. W. Samson, *Proc. Int. Conf. on Cloud Physics*, Toronto, 1968, p. 768.

94   E. G. Droessler, *Science*, **162** (1968), p. 287.

95   S. A. Changnon Jr in (*c*), p. 420; and in *Bull. Am. Met. Soc.*, **58** (1977), p. 20.

96   I. I. Burtsev in (*d*), p. 217; L. J. Battan in (*l*), p. 4.

97   E. K. Fedorov in (*a*), p. 397.

98 (*r*) D. Atlas, *Science*, **195** (1977), p. 139.

99   J. C. Thams *et al.*, *Veroeffentlichungen der Schweizerischen Meteorologischen Zentralanstalt*, **2** (1966); P. Schmid, *Proc. 5th Berkeley Symp. on Mathem. Statistics and Probability*, **5** (1967), p. 141; J. Neyman and E. Scott in (*c*), p. 455.

100   H. Byers in (*a*), p. 32; H. Grandoso and J. Iribarne, *Zeitschr. f. Angew. Math. und Phys.*, **14** (1963), p. 549.

101   D. Atlas in (*r*).

102   *Ibid.*; NHRE, *Final Report*, December 1976, Nat. Center for Atm. Res., Boulder, Colorado; *Preprints, 6th Conf. on Planned and Inadvertent Weather Modification*, Champaign-Urbana, Sect. VI.

103   B. Federer and A. Waldvogel in (*d*), p. 303.

104   D. Atlas in (*r*).

105   *Ibid.*; K. A. Browning and G. B. Foote, *Quart. Journ. Roy. Met. Soc.*, **102** (1976), p. 499.

106   J. Sedunov in (*n*), p. 8.

107   G. K. Sulakvelidze *et al.* in (*a*), pp. 413 and 428.

108   L. Katchurin *et al.*, in (*c*), p. 236.

109   W. Hitschfield in (*f*), p. 655; L. J. Battan in (*l*), p. 19.

110   G. K. Sulakvelidze *et al.* in (*a*), p. 421.

111   L. J. Battan in (*l*), p. 13.

112   *Ibid.*, pp. 12, 9 and 5.

113   L. Krastanov and K. Stanchev in (*c*), p. 239.

114   L. J. Battan in (*l*), pp. 5 and 9.

115   S. A. Changnon Jr in *Bull. Am. Met. Soc.*, **58** (1977), p. 27.

116   S. W. Borland and J. S. Snyder in (*f*), p. 686; D. Atlas in (*r*).

117   *NHRE Final Report*, December 1976, NCAR, Boulder, Colorado, p. 19.

118   L. J. Battan in (*l*), p. 4.

119   B. Federer *et al.*, *Journ. Weather Modif.*, **7** (1975), p. 177; B. Federer *et al.*, *Grossversuch IV: Design of a Randomized Hail Suppression Experiment using a Soviet Method*, Eidgen. *Hagelkommission Wiss. Mitteilungen*, No. 81, ETH Zurich, 1977; B. Federer, *Am. Met. Soc. Monograph*, **38** (1978), p. 215.

120   *Eidgen. Hagelkommission*, *Wiss. Mitteilungen*, No. 80, ETH Zurich, 1977, pp. 61 and 92.

121   Quoted by L. J. Battan in (*l*), p. 4.

122   *Eidgen. Hagelkommission, Wiss. Mitteilungen*, No. 80, p. 7.
123   D. M. Fuquay in (*a*), p. 604.
124   G. Dawson in (*a*), p. 598.
125   *Ibid.*, p. 601.
126   A short survey of relevant theories is given in G. Breuer,
       Luftionen und Gesundheit, *Naturw. Rundschau*, **27** (1974),
       p. 194; see also A. Krueger, *New Scientist*, 14 June 1973, p. 668.
127   D. Fuquay in (*a*), p. 607.
128   I. I. Gaivoronski *et al.* in (*c*), p. 267.
129   H. W. Kasemir in (*c*), p. 298; and in (*a*), p. 612.
130   L. J. Battan in (*l*), p. 6.
131   H. W. Kasemir in (*a*), p. 622.
132   H. Riehl, *Science*, **141** (1963), p. 1001; R. C. Gentry in (*a*),
       p. 500.
133   *Ibid.*, p. 497.
134   R. G. Fleagle *et al.* in (*q*), p. 6.
135   (*s*) W. O. Roberts in H. J. Taubenfeld (editor), *Weather
       Modification and the Law*, New York, 1968, p. 10; J. A.
       Crutchfield in (*b*), p. 209.
136   R. C. Gentry in (*a*), p. 507.
137   R. I. Sax in (*f*), p. 665; R. C. Gentry in (*a*), p. 499.
138   R. I. Sax in (*f*), p. 666.
139   W. M. Gray and W. M. Frank in (*k*), p. 192; W. M. Gray and
       W. M. Frank in (*d*), p. 425.
140   *New Scientist*, 30 August 1973, p. 503.
141   G. J. F. MacDonald in (*e*), p. 171.
142   R. Davies-Jones and E. Kessler in (*a*), p. 590.
143   T. J. Henderson and W. J. Carley, *Preprints, 3rd Conf. on
       Weather Modification*, Rapid City, 1972, Amer. Met. Soc.
       Boston, 1972, p. 333.
144   D. Davies in (*k*), p. 334.
145   Joanne Simpson and A. A. Dennis in (*a*), p. 277.
146   W. M. Gray and W. M. Frank in (*k*), p. 192.
147   E. K. Fedorov in (*a*), p. 394; I. W. Litvinov in (*j*), p. 28.
148   M. J. Sallinger and J. M. Gunn, *Nature*, **256** (1975), p. 369.
149   B. M. Gray, having studied relevant European, Chinese,
       Japanese and American literature sources, comes to the
       conclusion that east–west shifts in the pattern of the general
       circulation occurred in the past and produced warming and
       cooling at different times in various places in the northern
       hemisphere (*Weather*, **30**, p. 359).
150   M. Milankovitch, *Roy. Serb. Acad., Specl. Publ.*, **132** (1941);
       J. Gribbin, *New Scientist*, 30 September 1976, p. 688; N. Calder,
       New Scientist, 9 December 1976, p. 576.

151   J. B. Hays *et al.*, *Science*, **194** (1976), p. 688.
152   W. H. McCrea, *Nature*, **255** (1975), p. 697; G. E. Williams, *Earth and Planetary Science Letters*, **26** (1975), p. 487.
153   J. Smagorinsky in (*a*), p. 633.
154   J. Kennet and R. Thunell, *Science*, **187** (1975), p. 487.
155   F. A. Street and A. T. Grove, *Nature*, **261** (1976), p. 385.
156   A. T. Wilson, *Nature*, **201** (1964), p. 147; G. J. F. MacDonald in (*e*), p. 174.
157 (*t*) M. I. Budyko and K. Y. Vinnikov in W. Stumm (editor), *Global Chemical Cycles and their Alterations by Man*, Dahlem Konferenz, Berlin, 1977, p. 189; H. Flohn, *ibid.*, p. 207.
158   W. W. Kellog and S. H. Schneider, *Science*, **186** (1974), p. 1163; Y. K. Fedorov in (*a*), p. 400; M. I. Budyko and K. Y. Vinnikov in (*t*), p. 189.
159   S. A. Changnon Jr in (*f*), p. 654.
160   *Ibid.*, p. 653; R. R. Braham Jr in (*d*), p. 435.
161   B. Bolin, *Science*, **196** (1977), p. 613; G. M. Woodwell *et al.*, *Science*, **199** (1978), p. 141.
162   A. Lerman *et al.*, J. J. Morgan *et al.* and A. Nir *et al.*, Group Reports of the Dahlem Konferenz in (*t*), p. 275; G. Breuer, *Geht uns die Luft aus?*, Stuttgart, 1978. (English Edition, Cambridge University Press.)
163   H. Flohn in (*t*), p. 207.
164   M. I. Budyko and K. Y. Vinnikov in (*t*), p. 203.
165   E. K. Federov in (*a*), p. 400.
166   M. I. Budyko, *Proc. Symp. on Arctic Heat Budget*, Santa Monica, California, 1966, p. 3; H. Flohn in (*t*).
167 (*u*) For a summary see W. W. Kellog and S. H. Schneider, *Science*, **186** (1974), p. 1163.
168   M. I. L'vovich, *Priroda*, **3** (1978), p. 95 (*New Scientist*, 21 September 1978, p. 834). K. Aargaard and L. K. Coachman, *EOS*, **56** (1975), p. 484; The Soviet hydrologist T. W. Odrova points out that the installation of hydroelectric power stations on Siberian rivers might produce the opposite effect. Water in deep reservoirs remains cold at the bottom, and the temperature of water going through the turbines is around 4 °C. This compares with summer temperatures of around 20 °C in freely flowing rivers; it implies that less heat is transported northwards when dams are installed (*Priroda*, **6** (1977), p. 92; *New Scientist*, 18 August 1977, p. 411).
169   E. K. Fedorov in (*a*), p. 399.
170   P. M. Borrisov, *Isvestia Acad. Nauk. SSSR*, Ser. Geoph., **2** (1969), p. 47.
171   M. I. Budyko, quoted by S. A. Changnon Jr in (*b*), p. 161.

172   H. Flohn in (*t*).
173   W. W. Kellog and S. H. Schneider in (*u*), p. 1169.
174   *Ibid.*, p. 1170.

3 Problems and dangers

1   B. C. Farhar in (*k*), p. 552.
2   S. Krane and J. E. Haas in (*k*), p. 570; J. E. Haas in (*a*), p. 798.
3   *Ibid.*, p. 798.
4   I. Burton in (*b*), p. 68.
5   A. Meadows *et al.*, *Limits of Growth*.
6   M. Nicholson, *The Environmental Revolution*, Hodder and Stoughton, London, 1970.
7   G. Breuer, *Die Herausforderung, Energie fuer die Zukunft – Gefahren und Moeglichkeiten*, Munich, 1975, p. 210.
8   W. R. D. Sewell in (*b*), p. 40.
9   *Ibid.*, p. 42.
10   M. Holden Jr in (*b*), p. 306.
11   A. Morris in (*s*), p. 163; A. Morris was Pacific Gas & Electric's lawyer in this lawsuit.
12   D. E. Mann, *Bull. Am. Met. Soc.*, July 1968, p. 690.
13   R. D. Elliot in (*a*), p. 82; Elliot was the head of this company.
14   *Ibid.*, p. 84; W. R. D. Sewell in (*q*), p. 98.
15   B. C. Farhar in (*k*), p. 557.
16   A. M. Kahan *et al.* in (*d*), p. 31; G. W. Brier and G. T. Meltesen in (*d*), p. 181.
17   G. W. Brier, L. O. Grant and P. W. Mielke Jr in (*k*), p. 510.
18   L. O. Grant in (*a*), p. 315.
19   G. J. Mulvey and L. O. Grant in (*d*), p. 473.
20   G. W. Brier, L. O. Grant and P. W. Mielke Jr in (*c*), p. 439.
21   G. W. Brier *et al.* in (*k*), p. 512.
22   *Ibid.*, p. 513.
23   Joanne Simpson and A. A. Dennis in (*a*), p. 274.
24   J. E. Haas in (*a*), p. 801.
25   J. A. Crutchfield in (*b*), p. 198.
26   C. F. Cooper in (*b*), p. 116.
27   J. A. Crutchfield in (*b*), p. 192.
28   *Ibid.*, p. 196.
29   J. E. Haas *et al.* in (*d*), p. 591.
30   R. G. Fleagle *et al.* in (*o*), p. 39; C. F. Cooper in (*b*), p. 127; J. A. Crutchfield in (*b*), p. 195; W. R. D. Sewell in (*b*), p. 34.
31   J. E. Haas in (*a*), p. 788.
32   I. Burton in (*b*), p. 58; J. A. Crutchfield in (*b*), p. 197.
33   C. F. Cooper in (*b*), p. 114; I. Burton in (*b*), pp. 51 and 73.
34   J. A. Crutchfield in (*b*), p. 187; and in (*q*), p. 111.
35   R. D. Elliot in (*a*), p. 84.

36   *Ibid.*
37   W. R. D. Sewell in (*b*), p. 23.
38   *Ibid.*, p. 327.
39 (*v*) C. F. Cooper and W. C. Jolly, *Ecological Effects of Weather Modification*, University of Michigan, 1969, p. 61.
40   (*v*), p. 51.
41   *Ibid.*, p. 61.
42   Joanne Simpson in (*a*), p. 262.
43   J. A. Crutchfield in (*q*), p. 110.
44   C. F. Cooper and W. C. Jolly in (*v*), p. 2.
45   *Ibid.*, p. 91.
46   *Ibid.*, p. 123.
47   (*m*), pp. 146 and 150.
48   C. F. Cooper in (*b*), p. 102.
49   C. F. Cooper and W. C. Jolly in (*v*), pp. 113 and 115.
50   *Ibid.*, p. 83.
51   *Ibid.*, pp. 34 and 44.
52   *Ibid.*, p. 120.
53   J. A. Crutchfield in (*q*), p. 113.
54   C. F. Cooper and W. C. Jolly in (*v*), p. 74.
55   *Ibid.*, pp. 48 and 62.
56   *Ibid.*, pp. 7, 89 and 144.
57   *Bull. Ecol. Soc. Am.*, March 1966, p. 39.
58   C. F. Cooper in (*b*), p. 118; (*m*), p. 148.
59   R. B. Standler and B. Vonnegut, *Journ. Appl. Met.*, **11** (1972), p. 1388.
60   D. A. Klein and E. M. Molise in (*k*), p. 529.
61   C. S. Downie and R. A. Dirks in (*d*), p. 560; J. Wisnievski in (*d*), p. 536; F. P. Parungo *et al.* in (*d*), p. 505.
62   L. J. Battan in (*l*), pp. 5 and 7 and 10; G. K. Sulakvelidze in (*a*), p. 418.
63   F. Barnaby, *New Scientist*, 1 January 1976, p. 8.
64   C. F. Cooper in (*b*), p. 120; C. S. Downie and R. A. Dirks in (*d*); L. J. Battan in (*l*), pp. 5, 10.
65   R. W. Johnson in (*s*), p. 76.
66   R. W. Johnson in (*h*), Sect. IX, p. 9.
67   *Rep. Spec. Comm. on Weather Modification*, National Science Foundation, 66-3, Washington, 1966, p. 102.
68   R. S. Hunt in (*q*), p. 132.
69   *Ibid.*, p. 128.
70   M. B. Fiering *et al.*, *Decision and Institutional Aspects of Weather Modification*, US National Technical Information Service, PB 201 101, p. 3.
71   L. W. Weisbecker, *Technology of Winter Orographic Snowpack Augmentation in the Upper Colorado River Basin*, Stanford Res. Inst., EGU-1037, 1972; (*m*), p. 145.

72 R. D. Elliot in (*a*), p. 83; R. J. Davis in (*k*), p. 559.
73 *Ibid.*
74 R. J. Davis in (*a*), p. 770; R. D. Elliot in (*a*), p. 83; R. G. Fleagle *et al.* in (*o*), p. 44.
75 R. S. Hunt in (*q*), p. 125; R. J. Davis in (*a*), p. 769.
76 *Ibid.*, p. 769; R. S. Hunt in (*q*), p. 126.
77 M. Tribus in (*g*), p. 209.
78 R. W. Johnson in (*h*), Sect. IX, p. 14; H. J. Taubenfeld in W. R. D. Sewell (editor), *Human Dimensions of the Atmosphere*, NSF, US Govt Print. Off., Washington, 20402, 1968, p. 99.
79 Edith Brown-Weiss and W. H. Lambright in (*k*), p. 535.
80 R. G. Fleagle *et al.* in (*o*), pp. 64 and 80.
81 J. E. Haas in (*a*), p. 806.
82 *Ibid.*, p. 792.
83 *Ibid.*, p. 808.
84 R. D. Elliot in (*a*), p. 83.
85 M. Tribus in (*g*), p. 209.
86 Further examples can be found in J. E. Haas in (*a*), p. 788.
87 *Ibid.*, p. 807.
88 R. J. Davis in (*a*), p. 778.
89 *Ibid.*, p. 775 which also discusses Australian laws.
90 W. R. D. Sewell in (*q*), p. 101.
91 *Ibid.*, p. 101.
92 R. G. Fleagle *et al.* in (*o*), p. 39.
93 G. W. Brier in (*a*), p. 222.
94 R. S. Hunt in (*q*), p. 137.
95 W. O. Roberts in (*s*), p. 9.
96 R. A. Howard *et al.*, *Science*, **176** (1972), p. 1191; R. W. Kates, Letter to editor of *Science*, **177** (1972).
97 W. O. Roberts in (*s*), p. 10.
98 R. G. Fleagle *et al.* in (*o*), p. 45.
99 *Ibid.*, p. 48; Edith Brown-Weiss and W. H. Lambright in (*k*), p. 539.
100 J. C. Oppenheimer and W. H. Lambright, *Southern California Law Review*, **45** (2) (spring 1972), p. 570.
101 M. Holden in (*b*), p. 304.
102 Edith Brown-Weiss in *Survival*, Int. Inst. for Strategic Studies, March 1975.
103 Edith Brown-Weiss and W. H. Lambright in (*k*), p. 537.
104 R. G. Fleagle *et al.* in (*o*), p. 62.
105 *Ibid.*, p. 27.
106 *Science*, **184** (1974), p. 1061.
107 Edith Brown-Weiss and W. H. Lambright in (*k*), p. 537.
108 M. Holden in (*b*), p. 305.
109 Edith Brown-Weiss and W. H. Lambright in (*k*), p. 537.
110 *Ibid.*, p. 537.

168

111   M. Holden in (*b*), p. 280.
112   G. J. F. MacDonald in (*q*), p. 82; R. G. Fleagle *et al.* in (*o*), p. 64.
113   Edith Brown-Weiss and W. H. Lambright in (*k*), p. 537.
114   R. G. Fleagle *et al.* in (*o*), p. 64.
115   J. W. Canan in (*p*), p. 357.
116 (*w*) Rita Taubenfeld in H. J. Taubenfeld (editor), *Controlling the Weather*, New York, 1970, pp. 49 and 61.
117   J. B. Wiesner and H. F. York, *Scientific American*, October 1964.
118   R. G. Fleagle *et al.* in (*o*), p. 49.
119 (*x*) *Weather and Climate Modification – Problems and Prospects*, US Nat. Acad. Sci./Nat. Res. Council, 1350, Washington, 1966, p. 26; D. E. Mann in (*s*), p. 156.
120   G. J. F. MacDonald in (*e*).
121   J. W. Canan in (*p*), p. 357.
122   92nd Congress, 2nd Session, Senate's Resolution 218; full text in (*m*).
123   Edith Brown-Weiss in *Survival*, Int. Inst. for Strategic Studies, March 1975.
124   F. Barnaby, *New Scientist*, 1 January 1976, p. 7; the SIPRI is an institution created on recommendation from the Pugwash Conferences.
125   E. K. Fedorov in (*a*), p. 401; in recent years Professor Fedorov has made no public appearance, neither at international conferences nor in the Soviet Union; the reasons for this are not known; also it is unknown whether he has still influence in this field.
126   UN Document A/C 1/L675, 24 September 1974.
127   J. W. Canan in (*p*), p. 360.
128   (*x*); the emphasis has been placed by the author of this book.
129   E. K. Fedorov in (*a*), p. 400.
130   W. O. Roberts in (*s*), p. 17; see also T. F. Malone, *Science*, 19 May 1967, p. 900.
131   Edith Brown-Weiss in *Survival*, Int. Inst. for Strategic Studies, March 1975.
132   S. H. Schneider in *The Genesis Strategy*, Plenum Press, New York, 1976, p. 21.
133   A. Kiss in (*d*), p. 551.
134   A relevant proposal by the association called World Peace and World Justice is mentioned by R. J. Davis in (*a*), p. 785.
135   Rita Taubenfeld in (*w*), p. 92.
136   *Ibid.*
137   C. F. von Weizsäcker, *Bedingungen des Friedens*, Göttingen, 1964.

Glossary of technical terms

Cirrus clouds: White clouds at high altitudes, consisting of ice crystals; usually with filaments and wisps.

Cloud seeding: Introduction into clouds of a substance to stimulate precipitation; silver iodide is most widely used, the crystalline structure of which resembles that of frozen water.

Condensation nuclei: Particles in the atmosphere on which water vapour condenses and droplets form.

Convective clouds: Clouds generated by the rising of air over relatively warm ground.

Corona effect: Electrical charging (ionisation) of air near pointed objects or conductors.

Cumulonimbus clouds: Dense massive clouds, often associated with thunderstorms, appearing dark when viewed from below; towering to great altitudes and always precipitating.

Cumulus clouds: Rounded heaped clouds of cauliflower shape, generally associated with fair weather; depending on weather conditions they dissolve after a while, or develop into cumulonimbus clouds.

Dew point: The temperature at which air is saturated with water vapour; cooling of moist air below its dewpoint produces condensation (dew on plants).

Dynamic cloud seeding: Seeding of a cloud with the intent of increasing the internal processes that lead to its development and growth. The introduction of abundant numbers of icing nuclei causes rapid freezing of large numbers of supercooled cloud droplets, thereby liberating energy in the form of latent heat inside the cloud.

Ecology: A branch of biological science dealing with relations of living organisms to their surroundings, their habits, modes of life, etc.

Eye (of a hurricane): Almost calm, circular area of lowest pressure in the centre of a hurricane; it is surrounded by the 'eye wall' in which winds reach their highest intensity.

Föhn: Very dry warm wind coming down on the leeward side of a mountain range after having lost most of its moisture on the windward side. Brings fair weather, but is associated with harmful effects on the nervous system – causing severe migraine, nervous tension, deterioration in the state of many sick people, increased suicide and accident rates etc.

170

Hurricane: Violent large-scale revolving wind storm in the tropical Atlantic; Similar storms (typhoons) occur in the Pacific and Indian Oceans (where they are known as cyclones).

Inversion: Anomalous increasing of temperature with height in the atmosphere; usually temperatures decrease with height.

Icing nuclei: Particles in the atmosphere which induce freezing of supercooled water droplets.

Latent heat: Heat taken up, or given off, by a substance when it changes its physical state (solid, liquid, gaseous); on evaporation, water vapour takes up heat from the water surface (latent heat of evaporation), on condensation it warms the surrounding air (latent heat of condensation).

Mathematical model: Description of a physical process by a system of equations; electronic computers are widely used to provide numerical solutions. Such models facilitate the study of effects on the atmosphere produced by changes in the environment.

Orographic clouds: Clouds which form as a result of the shape (topography) of the ground; in particular when winds blow up the slope of a mountain.

Overseeding: Use of large amounts of seeding substance to produce great amounts of tiny ice crystals which are carried far afield by the winds; this generally reduces precipitation.

Ozone layer: Region of the upper atmosphere (around 30 km altitude) where the ultraviolet component of sunshine generates ozone (molecules comprising three oxygen atoms). The ozone layer shields the surface of the earth from harmful radiation.

Placebo: Mock medicine.

Randomisation: Precaution taken in the design of an experiment to ensure that statistical inferences to be drawn are incontestable.

Stratus clouds: Stratified cloud decks of amorphous shape, similar to fog.

Tornado: Violent, tubular wind vortex, of small diameter and short duration; occurs frequently in North America.

WMO (World Meteorological Organization): A specialised agency of the United Nations which co-ordinates scientific and technical work in meteorological services throughout the world.

Subject Index

Page numbers in italic type indicate reference to a figure or table

ARPA (Advanced Research Project
Agency), 145
'Agnes', US hurricane (1972), 91
agriculture: economic considerations of
weather modification on, 119–22;
effects of rainfall stimulation on,
67–8
aircraft: use of, for cloud seeding with
silver iodide, 46, 61
Arctic, 3; climate modification by
removal of sea icepack from, 106,
107, 108
atmosphere, upper layers of, xii

carbon dioxide, 129; and 'cold' fog
dissolution using liquid, 45; from
fossil fuels, and effects on climate,
104–5; *see also* dry ice
'Carla', US hurricane (1961), 7
'chaff', use of, for lightning prevention,
88–9
cirrus cloud, *24*, 117, 170
Cirrus Project, 33, 70, 144
'Cleo', US hurricane (1964), *140*, 140–1
climatic change: and effects of
continental drift on, 99–100;
proposals for future methods of,
106–8
climatic factors, and effects on weather,
3–4
cloud cap, *see* orographic cloud
'cloud mergers', 64
cloud seeding, *see* seeding, of clouds
clouds, 11; formation and development
of, 23–8; 'warm' clouds, seeding of,
65–6; *see also individual cloud types
and* seeding, of clouds
computing methods and models: for
seeding conditions, 54; of tornadoes,
95; of weather forecasts, 5–6, 96
condensation nuclei, *see* nuclei,
condensation

continental drift, and effects on climatic
change, 99–100
corona effect, and occurrence of
lightning, 88, 89, 170
cost–benefit calculations, *see* economic
considerations
cumulonimbus clouds, 27–8, *37*, *60*, 61,
170
cumulus clouds, 9, 18, 20, 58–65, 118,
170; development of, 23, *24*; dynamic
seeding of, 63–5; suitability of, for
seeding, *37*, 38; *see also* seeding, of
clouds

'Debbie', US hurricane (1969), and
modification experiments on, 92–3,
93
defence, cloud seeding as a method of,
70–5; *see also* warfare, environmental
depressions, *13*, 38, 68, 96–7
drought, 14; triggering of, xii, 65–6
dry ice, 129, 133; and 'cold' fog
dissolution, 45; and early cloud
seeding experiments with, 31, *36*
dynamic cloud seeding, 38, 63–5, 117,
170

ecological consequences, 126–30
economic considerations, 118–22
Edison Company of Southern
California, USA, seeding programme
of, viii
energy exchange, in weather systems,
13–14
European Centre for Medium Range
Weather Forecasts, England, 8

FIDO-systems (Fog, Intensive
Dispersal of), 40
fires, forest: use of cloud seeding for
combating, 68, 122
fog, and fog dispersal, viii, ix, 14, 18,

172

fog, and fog dispersal (*cont.*)
25, 39–47, 122, 125, 128; 'cold' fog,
dissolution of, 44–5; FIDO-systems,
40; increase of, by industrialisation,
104; 'warm' fog, dissolution of, 41–4
föhn (foehn), 38, 170
forecasting, of weather, 5–8, 95–6;
long-term, 100–1
forests, and effect on climate and
environment, 103; use of cloud
seeding for combating fires, 68, 122
France, fog dispersal at airports, ix
freons in atmosphere, and effects on
climate, 105
frontal cloud masses (depressions), *13*,
38, 68, 96–7

Geneva Arms Limitation Talks, xii
'Ginger', hurricane (1971), 92
greenhouse effects, 104, 105
ground generators: for hail prevention,
77, 78; and silver iodide cloud
seeding using, *39*, 46, 52, 54

hail prevention, viii, ix, 11, 16, 22,
114–15, 120, 128, 130; hail cannon,
75; legal aspects of, 134; rockets used
for, 46–7, 75–6, 82–6
heat, waste, and effect on climate, 105
helicopters, use of, in dissolving 'warm'
fog, 41–2
high pressure areas, modification of,
97–8
HIPLEX Project, 65, 117
human aspects, *see* social aspects
hurricanes, 12, 34, 148, 171; detection
of, by weather satellites, 7; energy of,
13; international problems of
modification of, 140–4; modification
experiments of, 90–5
hydroelectric stations, 129, 138–9;
economics of cloud seeding to
increase reservoir capacity, 69–70,
122
hydrological cycle, influence of
industrialisation on, 104

ice ages, and their causes, 98–103
icing nuclei, *see* nuclei, icing
India: hail damage in, 75; rainfall
enhancement experiments in, 22, 65
industrialisation, and effects on climate,
103–5
'Inez', hurricane (1966), 142

international aspects, of weather,
135–9, 151–7; and problems of
hurricane modification, 140–4; *see
also* warfare, environmental
Israel, rainfall stimulation in, ix,
16–17, 49, 68–9, 118, 136

Large Scale Experiments (I–IV),
Switzerland, on hail prevention, 75,
77, 78, 85–6
lead iodide, in rockets for hail
prevention, 77, 130
legal aspects, 130–5; and right of
ownership of weather, 135–9; *see also*
international aspects
lightning, 86–9, 115; 'chaff'
experiments for prevention of, 88–9;
energy of, *13*; Project Skyfire, 88
low pressure areas, modification of,
96–7

meteorology, as a possible future exact
science, 1–8
Milankovitch theory, 100, 102
military aspects, *see* warfare,
environmental

NHRE, *see next entry*
National Hail Research Experiment
(NHRE), 78, 81, 84, 117, 127
Naval Weapon Center, California,
USA, 70, 72; and hurricane
modification experiments, 92–3, *93*
Nile Blue Project, 145
nitrous oxide in atmosphere, and effects
on climate, 105
nuclei, condensation, 66, 170; and
formation of clouds, 25–6, 28
nuclei, icing, 20, 27, 30, 47, 81–2, 171;
production of, using silver iodide, 32;
seeding with, for 'cold' fog
dissolution, 44

oil-film method, for reducing storm
activity at sea, 94–5
Operation Popeye, 71
orographic cloud (cloud cap), 23–4,
171; problems of seeding, 115–16,
122, 123; seeding of, 38–9, 50–8, 62,
117, 118; snowfall increase, by
seeding of, 128
overseeding, 37, 39, 56–8, *58*, 171; *see
also* dynamic seeding
ownership, of weather, *see* legal aspects

tsunamis (flood waves), 148
Turboclair, 'warm' fog dissolution, 43–4
typhoons, see hurricanes

ultrasound, use of, for 'warm' fog dissolution, 44
United Nations Environmental Conference, Stockholm (1973), 151
urbanisation, and effects on climate, 103–5
urea, use of, in 'warm' fog dissolution, 42
USA, ix; river flow and effects of cloud seeding, 56; seeding of orographic cloud systems, 50–7; see also Vietnam
USSR, ix, 97–8; and fog dissolution, 45; hail prevention with rockets, 47, 77, 82–6, 130; and international aspects of weather modification,

147–50; lightning prevention using 'chaff' in, 89

Vietnam, xi, 42, 145, 148; cloud seeding in, 71–5
volcanic activity: and incidence of ice ages, 101; and effects on weather, 4

WMO, see World Meteorological Organization
warfare, environmental, and cloud seeding as a weapon, 70–5, 144–51; see also Vietnam
Whitetop Project, 49–50
waterspout, energy of, 13
Weather and Environmental Warfare, Treaty on, xii
weather systems, energy exchange in, 13–14
World Meteorological Organization (WMO), xiii, 22, 67, 82, 152, 171

Author Index

References in parenthesis indicate the actual reference number under which an author may be found in the text and notes

Aargaard, K., 108(168)
Artsybaschew, E. S., 68(73)
Atlas, D., 22(18), 77(98), 78(101–2),
  79(104), 81–2(105), 84(116)

Barnaby, F., 130(63), 149(124)
Battan, L. J., 61(50), 62(52), 66(66),
  77(96), 83(111–12), 84(114, 118),
  86(121), 89(130), 130(62, 64)
Bergeron, T., 27(23), 28(24), 35(3), 36,
  37, 38, 39
Berson, A., 26(21)
Bibilaschwili, N. S., 47(19)
Biodini, R., 50(29), 65(60)
Biswas, K. R., 65(62)
Bolin, B., 104(161)
Borland, S. W., 84(116)
Borrisov, P. M., 108(170)
Bowen, E. G., 68(74)
Braham, Jr., R. R., 104(160)
Breuer, G., 88(126), 112(7)
Brier, G. W., 19(13), 117(16–17, 20),
  118(21–2), 141(93)
Browning, K. A., 80, 81–2(105)
Brown-Weiss, Edith, 136(79), 143(99),
  144(102–3), 145(107, 109–10),
  146(113), 148(123), 151(131)
Budyko, M. I., 102(157), 103(158),
  105(164), 106(166), 108(171)
Buikov, M. V., 61(51), 68(72), 83
Burton, I., 110(4), 121(32), 121–2(33)
Burtsev, I. I., 77(96)
Byers, H. A., viii(3), 12(8), 35(1),
  65(63), 66(65), 70(78), 75(92), 78(100)

Calder, N., 100(150)
Canan, J. W., 73–4(88), 146(115),
  148(121), 149(127)
Carley, W. J., 95(143)
Changnon, Jr, S. A., 21(16), 49(27),

76–7(95), 84(115), 103(159), 104(160),
  108(171)
Charak, M. T., ix(6)
Chary, H. A., 41(7), 45(15)
Coachman, L. K., 108(168)
Cooper, A., 55(40)
Cooper, C. F., 120(26), 121(30),
  121–2(33), 124(39), 127(44–6, 48),
  128(49–52), 129(54–6), 130(58, 64)
Crutchfield, J. A., 75(90), 91(135),
  119(25), 120(27–8), 121(30, 32),
  122(34), 124(43), 129(53)

Davies, D., 96(144)
Davies-Jones, R., 95(142)
Davis, R. J., 133(72–4), 133–4(75),
  139(88–9), 152(134)
Dawson, G., 87(124–5)
Dennis, A. A., 13(19), 21(15), 58(45),
  60, 63, 64(56, 58), 65(62), 66(64, 67),
  67(70), 96(145), 118(23)
Dessens, H., 66
DiGuilian, M. T., ix(6)
Dirks, R. A., 67(69), 130(61, 64)
Downie, C. S., 67(69), 130(61, 64)
Droessler, E. G., 76(94)

Elliot, R. D., viii(3), 51, 52–3(31),
  53(32–3), 56(42), 68(74), 116(13–14),
  123(35–6), 133(72–4), 138(84)

Facy, L., ix(8), 43(10), 45(13)
Farhar, B. C., 109(1), 116(15)
Federer, B., 47(18), 79(103), 85(119), 86
Fedorov, E. K., ix(7), 62(53), 77(97),
  97–8(147), 103(158), 106(165),
  108(169), 149(125), 151(129)
Fiering, M. B., 132(70)
Findeisen, W., 27(23), 28
Fischer, G., 2(1)

176

88322